Monkey Business for Today's Phone Weary Office™

Hear / Speak / See No Evil and Take Friday Off!

Here's What a Few Noted Reviewers Have to Say

I highly recommend Chris Mullins' book for any office that is serious about telephone skills management and getting the most productivity out of office personnel. After 38 years in business, I know that this is a crucial element in building a successful business and maintaining a positive atmosphere in interoffice communications. Very few teach this much-needed training and Chris is a real professional with a track record of success with many companies throughout the US. Chris Mullins can make your office a productive winner.

– Pete Lillo, President, Kennedy-Lillo Associates, Pete Lillo & Associates, PetethePrinter.com

This book zeroes in on several "elephants in the room" either willfully ignored or mysteriously invisible to most owners, notably the handling of all the inbound calls to their office: prospective customers responding to advertising, referrals, present and past customers. Tens of thousands, more commonly, hundreds of thousands of dollars are lost every year in the typical business. While natural to insist "not in mine," if you aren't actively working on managing and optimizing inbound call handling, then it is. Long before the presentation by the owner, the road is paved or insurmountable obstacles created by all the "little things" that have occurred for the customer up to that point. This is an important book.

– Dan Kennedy, www.NoBSBooks.com

Businesses invest thousands in lead generation without realizing there is an enormous hole in the bucket: poorly trained and poorly managed customer service phone staff. With her meticulous attention to detail and superb coaching and scripting expertise, there is no one better than Chris Mullins to fix the problem and help your business build and maintain a profitable front line.

– Susan Berkeley, author of Speak to Influence: How to unlock the hidden power of your voice, *www.greatvoice.com*

While Chris is titling her book *Monkey Business*, it could just as well have been called *The Most Critical Tool in Your Business*. What she is putting in your hands is an exact system for growing your business from the trenches. The **10-POINT SCRIPT OUTLINE CHEAT SHEET** on page 34 is absolutely priceless and when used exactly as illustrated, it can't help but add additional volume to any business office. It's about time that somebody is teaching businesses that everyone in the office is really part of a sales team, starting with how the receptionist answers the phone when a customer or prospect calls.

– Bill Glazer, President, Glazer-Kennedy Insider's Circle, www.dankennedy.com

This book is a guide for *all* businesses. Here is a partial list of the types of clients Chris currently works with.

- franchises
- information marketing
- healthcare
- cosmetic surgeons
- podiatrists
- retail
- auto repair
- dentists

Editor: Deb Morrison
Production and Layout: DAM Graphics!
Book Title Creation: Pete Lillo, Pete Lillo & Associates
(www.PetethePrinter.com)

Mullins Media Group™ LLC
507 Greenfield Road
Peterborough, New Hampshire 03458

Chris Mullins, CEO of Mullins Media Group™ LLC, has helped hundreds of companies improve their use of the telephone through mystery call shopping, script review and staff coaching. Ms. Mullins is an expert at dramatically increasing appointments (sales) from your existing incoming calls. She is legendary for coaching all types of business owners and sales professionals to "No Excuses" peak productivity. Chris Mullins, with the clients' permission, listens to your phone calls to design a custom incoming call script for you that is easy to implement and instantly makes you more money! For more information on these and all other programs, please go to

www.MullinsMediaGroup.com

Monkey Business for Today's Phone Weary Office™

Hear / Speak / See No Evil and Take Friday Off!

Chris Mullins, *The Phone Sales Doctor™*

Disclaimer and How to Use This Guide

This book is designed to provide information about the subject matter covered. It is sold with the understanding that the publisher and authors are not hereby engaged in rendering legal, psychological or other professional services. If expert assistance is required, the services of a competent professional should be sought.

We are available for consultation on a professional basis.

Every effort has been made to make this book as complete and accurate as possible. However, there may be mistakes both typographical and in content. Therefore, this text should be used only as a general guide.

The purpose of this book is to educate and entertain. The authors and publisher shall have neither liability nor responsibility to any person or entity with respect to any loss or damage caused or alleged to be caused directly or indirectly by the information contained herein.

Who Should Use This Book?

Some of the professionals that will benefit from this book are attorneys, accountants, dentists, opthamologists, veterinarians, automotive repair, hearing improvement businesses, real estate, tax preparers and health care providers. Consultants and coaches will also benefit from reading this guide.

- The back page of each chapter of this guide is an "AHA" notes page. Make this your book your own!
- Fax back your notes to Chris Mullins at 603-924-5770 if you would like discussion and feedback.
- Do the exercises inside this book.
- For additional resources on this book go to www.UniversityOfPhoneProfessionals.com and www.GreatBottomLine.com/fixmyphones.

Contents

Introduction

Who Am I, and Why Would You Want to Listen to Chris Mullins?

"We have met the enemy and he is us."
Walt Kelly

From Chris Mullins

Hello, this is Chris Mullins, The Phone Sales Doctor™. My style is pretty informal and I try to seem like I'm talking right to you. I'll be addressing the business owner, the receptionist, the front desk expert, the individual primarily answering the phone, the administrative assistant, the office manager, the sales team and customer service.

Make a commitment to set up a system as far as a schedule at your business where following each lesson you read in this book you include discussion about it in your team meeting.

I like to call the meeting Sales Drills™ and eventually you'll call them sales drills because we really want to have the mindset of a sales culture in your business.

While reading this book, make a list of all the things that make perfect sense to you. These are what I call Side Notes™. Everything that I'm going to discuss with you is important and you need to apply all of it, but when you're reading, sometimes, all of a sudden, you may hear something that sounds like a bell going off in your head. These are things that really kind of make you go, "Aha!" or excite you.

Make a list of those spots first and then make a commitment to put a step in place — an action step — which achieves that list, because those are the things that you'll probably get accomplished the quickest and have the best feelings about to get your confidence up when it comes to tackling anything else on the list.

In addition, give yourself a deadline. You can't just make an action list of things that you're going to fix. You've got to have what you're going to fix and how you're going to fix it, but you also need to put a deadline on it to help you commit to doing it. With that, let's begin.

The Most Critical Tool in the Office

This book is about the crucial but often missing link between our expensive professional skills and actually closing sales — the **telephone**.

Here's the sad truth: You can be highly skilled, beautifully equipped, have top-notch staff, run killer ads and second-to-none marketing campaigns, and STILL miss out on actually doing business with the customer… *if people aren't trained to use the phone correctly.* If the phone isn't working right, you'll have:

- far fewer appointments getting scheduled
- far fewer appointments showing up
- more surprise no-shows happening
- overall far fewer sales going forward, and
- greatly reduced sales on your investment

That first phone interaction after all your correct marketing, screening and qualifying can literally be either the deal maker or the deal breaker. As can the correct followup.

By the way, if you register at www.GreatBottomLine.com you'll get **FREE Money-Making Telephone Skills Reports** that will boost your staff's results right now with this critical business tool.

PS — There is also an important **FREE GIFT OFFER** from Chris Mullins on page 161 of this book. This gift can further boost your inbound phone skills and is an additional way to continue our relationship beyond the bounds of this book.

Chapter *One*

Fixing the Mindset of the Receptionist

I'd like to start this chapter by saying that without the receptionist, the front desk expert, you have no business. I think we can all agree that you're in business because you want to do good, you want to help service others with your products and services. However, this is a business and the mindset of the owner is critical with regards to the bottom line, but even more critical is the "sales mindset" of the receptionist. You can love what you do, still be a sales expert and still want to improve the bottom line of the business.

Maybe a Chris Mullins True Personal Story From the Good Old Days Will Help

My first real job was working as an inside salesperson for the classified advertising department of a large seven-day-a-week newspaper.

This job was a very big deal to me because it was the major newspaper in the town I grew up in. Whenever I drove by the *Lawrence Eagle Tribune* in North Andover, Mass., it seemed so big to me. Wouldn't it be great to work there someday?

I got the job and was there for several months and I thought I was incredible, the best — I really felt very strongly that I finally understood many of the important points of the job.

The big winner for me was how I handled the customers. I was unstoppable! No one could match the service I gave to each and every customer that called the classified advertising department. I felt very good about what I did each day.

Chris' BIG Thought

Business owners, pay close attention to the above message. In many cases this is how your team feels because you haven't taught them any different. Phil, my boss, didn't teach me either, or at least as often as he should have. Therefore, I thought my job was to do everything possible to make the customer happy, which to me meant *save them money* and keep them laughing and talking.

How do you measure up? Are you creating automatic production machines with the right language, Sales Talk™ and ongoing coaching in your business? Or, are you sending the wrong message just like Phil did? You've got to continually remind, police, support and coach your entire team.

The Rest of My Story...

Phil called me into his office and basically told me what a bad job I was doing. I was embarrassed and became very angry because I honestly had no idea. Phil's concern was that I was on the phone too long with customers, talking up a storm, not controlling the conversation and saving them money on their classified ads.

After that talk I really disliked the newspaper and wasn't happy at my job anymore because I really believed that the paper was out to steal from the customer.

As time passed and I attended various newspaper association sales conferences, I quickly learned that Phil was right. I was actually doing a disservice by trying to save advertisers money, because I wasn't giving them the opportunity to write the best story ever about what they wanted to sell, which meant it would get better, more qualified readership and they would have more opportunity to sell.

I realized that the longer I was on the phone with one customer the more I was hurting others waiting to be taken care of, that were on hold or simply hung up. I learned the hard way, but I did learn and that was the beginning of my true understanding of sales, the sales presentation and selling with integrity. I learned that customer service *is* sales. Sales *is* customer service. Selling to the customer's specific need *is* customer service.

The "S" Word (Sales)

Make life easy for yourself — save time and energy — and start from the beginning. Hire people that can sell or will be okay with selling, meaning, not offended by it. It's more important to hire people that understand sales, the bottom line, sales targets, scripts, goals and business than it is to hire those experienced in your business.

It's time to yell and scream at the top of your lungs the "S" word. Say it right now!

SALES
SALES
SALES

Put up "What did you sell today?" signs (everybody should have those everywhere, except, of course, where your customers will see them).

In fact, you (the owner/manager) and each team member must talk sales (Sales Talk™) everyday, no exceptions!

I'll begin by saying I LOVE SALES and I believe in sales and teach sales. I help folks like you identify sales moments. Think about it for a moment. We're all born salespeople, naturally trying to convince others of what we believe to be true and of what we want, trying to change beliefs, and in many cases we don't even realize it. Everything we do in our life involves sales. In fact, sales should be taught early on in junior high school, it would make life so much easier for students and it would make them better applicants in the job world which would help us all.

Sales is training for life, from trying to convince your partner of some new high-priced gizmo you want to buy, to convincing your daughter's teacher that even though she may have Attention Deficit Disorder you fully expect the team of teachers to work just as hard for her future success as they do with all other students, to convincing the store manager of a retail store that you expect a complete refund, etc.

There are bad apples and good apples out there, bad salespeople that lie, who don't care about anything they sell or the people they sell to. Then there's you and me; we love what we do and believe in it, but at the end of the day we know we must sell, increase our herd of prospective customers and retain our established customers or we're out of business. Basically, as long as you do your homework and match the right products and services to your interested prospects and established customers, you're golden. That's sales and, of course, you never give up. In fact, if you don't continually offer your prospects and established customers more products and services, whether they ask for them or not, you're doing them a disservice. They deserve to know what you have available and they deserve the opportunity to believe that they're worthy of them, so give them a chance!

Self-Motivation & Self-Discipline: The Discovery Mindset

As I've mentioned in the past, the common denominator that I've found and read about for folks that remain at the top — whether professionally or personally, whether top senior executives or frontline staff members — are those that continually educate, challenge and push themselves, the receptionist and the entire team to learn something new every day.

In addition, the team that remains at the top isn't just dedicated to learning something new, but they're committed and they're self-disciplined enough to apply what they've learned and go from there. They take action!

I'm not talking about the people that always score the highest on tests or college exams, I'm talking about everybody and anybody; you don't need a high IQ to develop your own personal system for being self-motivated and self-disciplined.

Some of you may say, "Every day?" Yes! Every day. You can learn so much about yourself AND those around you, every day.

There are lessons in everything we do and say, all that we hear and see others do. However, unless you, the business owner/manager, are in the mindset of continually learning over and over again, asking yourself the question, "How can I apply what I just heard or read to my own situation?" you will not be in the mindset of discovery and neither will your team.

One of the easiest and quickest ways to get your team into this discovery mindset (which is, of course, a new habit that you'll need to develop) is to simply encourage them to ask questions.

Ask questions of those around you. Many times we hear things that others say that we don't get, we don't understand, but we just push it aside; sometimes we even try to look like we get it.

Usually this happens because we're either intimidated ("They'll think I'm stupid") or we may be thinking, "I don't see the value in understanding what I just heard, it doesn't apply to me." Here are just a few ideas on how you can begin to get yourself and your team into the mindset of discovery.

POWER NUGGET™

While this chapter is about fixing the mindset of the receptionist, my writing includes the owner and ALL team members

1) Read. Read for fun. Read to understand more about your profession. Read to understand more about your customer's emotional buying behaviors. Read to learn about everything else.

2) Ask questions. Get into the habit of asking more questions of those around you, but don't forget you must also LISTEN to the answers.

3) Join a book club. Or start your own book club at the office for the staff to read that will help them work on their sales mindset, the business bottom line, their customers and customer service. This is a great way to get comfortable with others. Put yourself and your team in a situation where you

Resource: In fact, you can ask Chris Mullins questions by going to www.GreatBottomLine.com.

will read, discuss books, learn about yourself and other people at the same time.

4) Join Toastmasters. Get involved with an organized professional group or club that will stimulate you to step out of your comfort zone, learn and grow. This is an excellent resource I recommend for developing great speaking skills and presentation skills, both of which help the bottom line. (Go to www.toastmasters.org.)

5) Get a journal. Journal writing for professional and personal discoveries is like an ongoing, do-it-yourself, self-development training course. Write down your thoughts, feelings, discoveries, lessons and daily successes. Members of my Phone Success Training Program have either Sales Success Journals™ (SSJ) or Personal Success Journals™ (PSJ) — some have both. Find out more about these at www.GreatBottomLine.com.

6) Review your journal often — daily, weekly and monthly.

7) Goals. Set goals for yourself, put together a step-by-step action plan, add a start date, a deadline and be prepared to tweak along the way. Don't give up! Just do it!

Chris' thought-provoking reminder to ALL owners and team members in your business: The idea is to be sure that you're continually working *on* your business verses *in* your business (I learned this from www.DanKennedy.com).

Customer Service IS Sales

As the business owner, the leader, you have to get this. Then you need to quickly make sure that your receptionist and the entire team gets it. They simply have to be comfortable with the "S" word. In fact, you should all sing the word SALES over and over again at the top of your lungs, daily.

Understanding how true this really is that customer service is sales is the absolute quickest, easiest way to influence your team into becoming strong salespeople with integrity. Many clients share with me that they can't get Mary to sell because she says, "I'm not a salesperson. You didn't hire me to be a salesperson. I'm the receptionist, I just answer the phone."

Customer service taught correctly will get you the best of both worlds, but you do have to say the S word — S A L E S. Did you know...

1) Customers hear 50% of what you say and 100% of how you feel.

2) Customers are usually intimidated and they'll "yes" you when they don't really understand what you're saying to them.

3) Assume nothing.

4) Understand the importance of returning messages, e-mail, telephone and fax. Do what you say you'll do!

5) Learn how to listen to your customers.

Make the decision for the customer — that's what they want. Remember that your goal is to be of service to your customers and prospects. People need to be told what to do for a variety of reasons, but at the top would be the lack of confidence in their own decision-making ability. They may lack self-esteem, self-confidence and have a history of making the wrong decision.

So, Where Does the Receptionist Come In?

Your phone rings and the caller says, "Are you taking new customers?"

It doesn't matter why they call. If they called, they want to do business with you, now, and you must grab them by the shirt collar and pull them through the phone. Guide them and tell them what decision to make (which is, for example, to make an appointment). An easy way to do this is to say...

"Terrific! Mrs. Smith, you made the right decision calling ABC today. My name is Chris and it's nice to meet you. I have an opportunity for you to meet personally with ABC today at 10:00 am."

Your job, purpose, goal and focus with each and every call is to *ask for the appointment*. In many cases you'll ask three times in one conversation. This is how you make the decision for the customer.

You're Either In or Out! First Impressions

Understand that it starts with the receptionist. After listening to Zig Ziglar (an internationally renowned speaker and authority on high-level performance who offers easy-to-apply methods, techniques, exercises and tips in any career or endeavor), one of the very important things he shared was the GREETING, meaning how we greet others when they ask "How are you?" or

"How are things?"

The response that you want to remind, coach and lead by example with is, "Better than good!" — no matter how you're feeling. In fact, the "better than good" mindset should start internally, team member to team member.

Imagine if you initiated an internal campaign called Me First™ to check yourself out with the very focus being the greeting during all points of contact at all levels, internally and externally. That's what I would call the Wow Factor™! The more focus you place on your entire team starting with the receptionist regarding sales, internal customer service (owner to team member, team member to team member) and communication, the faster you'll improve your bottom line.

The really big question you want to ponder is how would that impact the bottom line with your customers? Remember, the phone is the first point of contact your established and prospective customers experience with your business, all pointing to the receptionist.

It starts with you — the owner. You're in a position of influence, and if anyone can make this happen, you can. Hold yourself accountable daily with your own sales mindset and internal communication skills, and your team will do the same which will automatically improve the bottom line.

Exercise: To further explore this concept, e-mail your thoughts and ideas to assistant@mullinsmediagroup.com and put MONKEY BUSINESS in the subject line.

Communicating With Customers

Every moment with a customer is a *sales moment*. When you have either a happy or unhappy customer, you must always be thinking about sales. Here's an example of a sales moment that wasn't initially thought of that way from one of my students enrolled in our Phone Success Program™, which happened to be for dentists.

> Chris,
>
> A customer just cancelled her maintenance visit and when I asked her in a disappointed tone the reason she said it was none of my business and why was I being so nosy? I explained but then she hung up. Should I call back tomorrow? She's usually a nice person.
>
> Linda

Hi Linda,

To answer your question, your customer may have been cancelling for a serious personal reason which probably caught her off guard when you asked why. Remember, it's not personal!

Don't stop asking why to others in a sincere "we're concerned" tone over the phone or in person. In person, you can lean forward, use eye contact and gently touch the person's shoulder. Over the phone, you'll simply use your voice, your tone, your words, appropriate pausing and exceptional listening skills.

I would need a little more information from you to give you a really good critique, Linda. You would need to tell me exactly what you said. When someone is upset at what we've done, don't bother explaining. Just listen and then apologize. In this case, yes, I would call back the next day and say…

"Mrs. X, this is Linda from Dr. ABC's office. Do you have a moment?"

Listen to her answer. If yes, then continue.

"Mrs. X, I was concerned after we spoke yesterday on the phone. I apologize if I upset you in any way."

Then PAUSE and wait for her reply. From there on, you'll know what to do.

Or you could immediately send a card in the mail; not a business card, but a real card, one that's not serious, but perhaps a cheer-you-up type of card.

Don't talk about the cancellation, just the fact that your customer seemed to be upset and that you're concerned.

Chris

It's important to help your entire staff, not just the front desk expert, understand what sales is really about — it's about customer service, it's upselling and it's sharing all your services and products with your prospects and customers.

POWER NUGGET™

Treat prospective customers like established customers and watch how fast you close your next sale

Where's Your Sense of Urgency?

If your numbers aren't where they should be today, right now — then find out why! Look at your sales tracking reports in your call measurement program provided by each associate in your business, by day, by hour, by month, by product and service. Compare them to the previous year, previous quarter, previous month, previous week, previous hour. Put your sales numbers under a microscope.

Team Exercise: Here's a quick example of the types of questions you and your team should be asking, but more importantly, you and your team must know the answers.

1) What's a Tuesday worth?
2) How long does it take to close a sale?
3) How much time is spent on scheduling a new customer?
4) Which outbound voice message scripts get the best results?
5) Which inbound phone slips (scripts) get the best results?
6) Which direct mail program gets the best response?
7) Which e-mail messages get the best response?
8) What's the best time to leave a message?
9) How many times did the phone ring today? (For more information on how to automatically track and measure your incoming calls, go to www.Great BottomLine.com.)
10) How many new appointments did you schedule today?
11) How many new appointment opportunities did you get today?
12) How many inbound calls does it take to close an appointment?
13) How many voice mail messages were left today?

Now, I really disliked putting the last point in here, because this just shouldn't happen at all. No voice messages, only live people, should answer the phone, however, you haven't yet read all of this book, so for now, track everything, follow every single lead you get and know at each moment where it is in the sales process, who's touching it now?

Back to the exercise. You'll be shocked at what you'll uncover. Here's a quick tip based purely on behavior and habits.

Resource: Go to www.GreatBottomLine.com for more information on our Call Measurement Recording Program.

Chris' BIG Thought

Many salespeople currently sell due to habit. If they're used to selling 20 appointments a day and no one is challenging them, that's exactly what they'll sell. It's all about *habits* and *behaviors*! Your confidence and comfort zone, of course, are part of the equation as well.

When Things Go Wrong

When things go wrong, as they sometimes will,
When the road you're trudging seems all up hill,
When the finds are low and the debts are high,
And you want to smile, but you have to sigh,
When care is pressing you down a bit,
Rest, if you must, but don't you quit.
Life is queer with its twists and turns,
As every one of us sometimes learns,
And many a failure turns about
When he might have won had he stuck it out;
So don't give up, though the pace seems slow –
For you may succeed with another blow.
Often the goal is nearer than
It seems to a faint and faltering man,
Often the struggler has given up,
When he might have captured the victor's cup.
And he learned too late, when the night slipped down,
How close he was to the golden crown.
Success is failure, turned inside out –
The silver tint of the clouds of doubt –
And you never can tell how close you are,
It may be near when it seems afar;
So stick to the fight when you're hardest hit –
It's when things seem worst that you mustn't quit!

– *author unknown*

"You're NOT Your Customer" Exercise for Staff

The purpose of the exercise below (which I received from a colleague) is to help the receptionist and all team members better understand their customers by putting themselves in the customer's shoes. To be reminded of what it's like to be a customer. To keep top of mind, on a daily basis, your "purpose." Many times I've recommended this type of exercise to clients, students and members of our Phone Success Program™.

Team Exercise: For one week team members should take the list below and keep track of their own personal experiences with customer service at the places they frequent over seven days — food shopping, online shopping, restaurants, movies, clothes shopping, banks, call centers, using the telephone to get personal projects and errands done, etc. Next, associates should share with the group how they would translate the experiences they had to the job they do every day.

1) How did you feel?

2) Did you feel welcome?

3) Did you feel rushed?

4) Were you made to feel as if you were the only person in the store?

5) Was your sales associate helpful?

6) If not, did they find someone who could help you easily?

7) Did your sales/customer service associate know their product?

8) Did you yourself discover the answers to your questions?

9) Did they introduce themselves and approach you well?

10) Did they help you buy, or did they let you buy?

11) Did you feel a positive control from the sales/customer service associate?

12) Did they add on sales?

13) Was their help useful?

14) Was it what you expected, less than or more than?

E-mail your results to assistant@mullinsmediagroup.com to get feedback from Chris.

Resource: To find out more about our Phone Success Program™, go to www.GreatBottom Line.com.

I once did a teleseminar for one of my private clients on attitude, smiling and mindset. I thought I'd share a few of the highlights with you.

1) View yourself as a Sales Business Owner (SBO) — the expert of the telephone.
2) Understand that your purpose is **sales**; at the end of the day that's all that really matters. What did you sell today? How many appointments did you sell?
3) Know specifically what works for you and do more of it. Know your strengths. Craft your presentation, your script, your message to each prospective and established customer, tweak where necessary until you have the secret presentation that you know works. Bulldoze through life's challenges!
4) Sharpen your axe, whether you're a seasoned professional or a new recruit. Always compete with yourself by asking, "How can I tweak that just a little more to save more time, sell more, cross-sell, upsell, shorten the sales cycle, get more referrals or manage my customer accounts and time more efficiently?"
5) Examine and develop ways for you — the sales professional (and celebrity customer sales coordinator) — to stand out amongst all the rest at your business, in your industry and more importantly from every other business out there that your clients and prospective clients communicate with day in and day out. *Smile, make the call and sell something!*

Three Words to Keep Top of Mind: Relax, Complacent, Readiness

Webster's Dictionary defines...

RELAX – *to unwind, loosen up, to rest*

The sales, customer service or business environment isn't for relaxing. The idea is to have a sales team full of energy, love what they do, but know and believe at the end of the day it's about sales and scheduling appointments. The receptionist, along with the rest of the team, is excited and full of ideas;

a team that believes in themselves as individuals and as a team. If you relax while playing the sales game, you'll take your eye off the ball, you'll slip up, miss a beat and you won't have both eyes on your focus and your purpose. At the end of the day all that really matters is... What did you sell?

COMPLACENT – *satisfied, self-satisfied, smug, unworried, content, self-righteous*

Be very careful, my students hear me speak of this often. Complacency is a life struggle for many professionals; a struggle to be able to quickly identify when complacency decides to show its ugly head. Don't be fooled; it shows up when you least expect it.

If you don't allow yourself to *relax* while playing the sales game, you'll be able to combat complacency. You'll always be focused and remain sharp.

READINESS – *willingness, gameness, eagerness, promptness, speediness*

Be ready the moment you drive into your office parking lot. Teach and guide your team to be ready for business daily.

The idea is to shake off everything that happened before you got into the office. It's game time! A great way to get ready is to have a "sales huddle" first thing every morning, five days a week, without fail. The owner/manager and all team members should do this to get ready for the day and have each team member share very quickly and concisely, in 10 minutes or less...

1) What they sold yesterday.

2) What they plan on selling today.

3) What they're forecasting for sales (appointments and beyond) for the week.

4) How they'll do it.

Keep the tone businesslike, positive and motivational. It's a sales huddle and everyone stands. They're charged up with a clear focus on the (daily) goal; their purpose — sales targets!

So, You Think You're a Salesperson?

Here's what *doesn't* make a good salesperson. Instead, you've become an excellent conversationalist. Just because you like the business you work for, you have a pleasant-sounding voice and you're courteous to the prospective customer, it doesn't make you a salesperson.

What's most important and missing from the above is, are you a CLOSER? Do you understand that your job is to sell appointments? Can you close appointments? Or do you just take no for an answer and say, "Call us when you're ready?"

You've got to get past all the customer service-type attitude stuff like, "I'll be nice and they will come or they'll call us when they're ready." This isn't the case; your customers and prospective customers want to be sold, that's why they called you. They want you to tell them what to do. They want you to take them by the hand and make the decision easy for them.

POWER NUGGET™

Major owner mistake

Again, at the end of the day all that really matters is how many more appointments did you sell today than yesterday?

Now that you realize you need help in sales, ask for it. Get sales coaching from the office manager, other top-performing individuals in your office or even the manager. Remember, the manager has to understand sales and how to sell. This is advanced selling and we teach it at www.GreatBottomLine.com.

An excellent way to coach the front desk person, owner and team members on how to focus on closing sales is to record calls and presentations, then listen to and critique them. This is the most important step and piece of information that I use in all of my training with owners and staff. You can do this, too.

Side Note: Hire individuals to work in your business that focus on sales — that's it. Someone to protect the sales process, all the steps involved and will be the point of contact (they must be well versed in presentation skills).

Resource: For more information on our Automatic Hiring System™, e-mail assistant@MullinsMediaGroup.com and put HIRING SYSTEM in the subject line, or go to www.GreatBottomLine.com.

AHA Notes:

Chapter *Two*

Selling Appointments

Rule # 1 – Use Your Phone Script

Rule # 2 – Ask for the Appointment

Rule # 3 – Repeat

You've got to remember that prospects are calling your office because they want to schedule an appointment, they want to do business with you — there's no other reason. Sure, they're consumers; they want to do their due diligence, their homework, before actually committing to the appointment, to giving you their business. So for many that means once they finally get around to calling you they will ask questions like, are you taking new customers, how much is it for the widget, what kind of financing do you have, what are your hours, etc.

Understand that this is exactly what you do when you're thinking about spending your money — whether you call an office or retail store or a spa, you're doing your homework.

The important point to note is that they didn't just decide today at 2:00 pm for the very first time to call you and ask these questions. They've been thinking about making this call for some time, but today at 2:00 pm they finally tipped over and set aside time to pick up the phone to call you and ask questions. This is your window of opportunity, you must grab them, hang on to them and, as mentioned earlier, pull them through the phone because they are in the right emotional state of mind to be scheduled for an appointment.

POWER NUGGET™

In order for the team to think like Sales Business Owners, the owner must treat them as such

So, you can say the wrong thing to make them decide that they don't need to do this now or you can say the right thing and convince them by saying "you made the right decision calling today" and schedule them for an appointment.

You must first understand that you're in the business of selling appointments regardless of how much you want to help people. This is the biggest hurdle that you must overcome. Once you and your staff understand that this is all about sales, the rest is easy, but everything you do in your business links back to the sales mindset of selling appointments. Therefore, the single most important thing to ask each other at the end of every day is "what did you sell today?"

The manager and each team member must think of themselves as business owners. The person that handles the telephone is the business owner of that

Resource: Go to www.UniversityOfPhoneProfessionals.com for script examples.

piece of the business, so the owner has outsourced his need for a telephone expert to this individual. Obviously, in order for the team to think like "sales business owners" the owner must treat them as such.

Five Secret Traits of Top-Performing Sales Business Owners (SBO)

Always Selling — They Never Stop

I recently heard a good analogy from one of my clients: "You're an actor; you're performing always, no matter what type of sales call you're on." When you go to a play, you expect the play to be perfect, even though the actors are doing the same play night after night and are exhausted. You still expect it to be perfect, you don't want to hear excuses that they may be tired. Same thing in sales, you're always selling and each customer prospect you speak to deserves the same experience, assertiveness and knowledge as the one before.

They Want It

They really want to sell, and anytime they have a challenge with how they sell, they really want to fix it. This has a lot to do with what motivates you. Knowing what motivates you deep down inside will open many doors for you (it's not money, by the way).

Relaxed and Comfortable

They focus on one call at a time. They engage the customer in conversation. They know sales is a process.

Never Make Excuses

They believe as individual sales business owners they're in complete control of their book of business, they take full responsibility and hold themselves accountable for all aspects of it.

They Love Sales

They absolutely love to sell — period! If you feel like you're doing some of the things we've talked about and that you're trying, that's good. It's a beginning and one action step usually leads to the next. However, I'd like to suggest that you take more risks. Go all the way. Instead of just putting your toe in the water, jump in all the way. You'll never know if you went too far until you've gone too far. One thing's for sure, you'll learn something new!

Learn From Sales Hot Shot™ Club Members

Here are three ideas from my Phone Sales students. Let's say you get new customers into your business by scheduling appointments.

1) Every single call is a sales, marketing and customer service sales moment — a "Wow Opportunity."
2) Tell the customer that you have an appointment for them rather than "Can I put you on a tentative schedule?" or "If you want you can come in this afternoon." Instead, try this: "Loretta, that (fill in the blank) doesn't sound good, I have an opening for you today at 10:00." Then PAUSE and wait for a response.
3) Get your 3 x 5 lime green psychological trigger card and set a daily goal for the number of appointments that you've decided to close for the day.

A great way to do this is to tally up the number of appointments you secured before you go home at the end of the day. If the number is five, then get your 3 x 5 card and make a decision of how many appointments you'll get tomorrow.

Let's say you want six for tomorrow. Write it on the card. Leave one card at your phone, another in the bathroom on the wall, tape another on the dash of your car, bring one home and tape it to your kitchen cabinet and, finally, bring another one to bed with you. Look at it before you go to sleep; close your eyes and say 10 times to yourself, "I will get six appointments." In the morning when you get up, before you get out of bed, look at the daily goal card that's on your night stand. Do this daily and you will win!

Six Tips for Goal Setting

In order to sell appointments and keep that focus top of mind you must have a plan, you can't just say, "I'm going to sell more appointments today than yesterday." Get your Sales Success Journal™ (SSJ) out and follow the six guidelines below.

1) Have a plan. Decide how many appointments you're going to sell today. Pick the number and write it on your 3 x 5 lime green index cards with a big black thick marker.
2) Be prepared to scrap the plan. If you find that your plan didn't work, make another one and move on.
3) Push yourself mentally like you're working towards the most important thing of your life. Expect a lot of yourself.
4) Do what you hate but know you have to do. This is an excellent way to learn and grow fast by putting yourself in uncomfortable situations on purpose.
5) Give everything you do day in and day out a goal. Compete with yourself. This will help you to get in the habit of pushing yourself. You can't just focus on changing behaviors and habits at the office. Start with small things outside the office.
6) Train your brain. Convince your brain through positive thinking and visualization. You'll be surprised at what you can accomplish when you say you can. See yourself writing your individual goal for scheduling new appointments on your 3 x 5 lime green index card, see yourself following your script — calmly, easily and effortlessly with complete enjoyment.

All or Nothing!

This is an ongoing behavior and habit you must accept. If you can't be committed to the telephone and can't fully embrace it as the most critical tool in your business, then you shouldn't be in that position. It's not fair to the business, the customers, the rest of the team or yourself. Think about it, athletes don't just train for the upcoming race — they train all year long.

1) Know what your "Individual Success Formula" is for your sport (the telephone). What does it take for you to improve your appointment scheduling closure ratio? Do you have to write in your SSJ™ (Sales Success Journal) everyday, six times a day? Do you have to record your own inbound phone calls and ask team members to critique them for you?
2) Duplicate each success step that you know works and include it in your personal life, too.

3) Practice a lot! Ask to be critiqued by the rest of the team in your weekly PAS (Problem, Agitate, Solution) — a very big focus with my students) sales drills.
4) Pay attention! Notice all the details in your successes, your actions, your new disciplines, your challenges. Use your SSJ™ and/or your PSJ™ (Personal Success Journal) daily. Write down everything — all your actions and thoughts.

The most important thing to understand is that everything you do in your life, not just at the office, affects your outcome, the final results to your goal. Here's an example of areas you'd want to consider for your journal: your emotions, what you ate, how you slept, exercise, what you read, your inner circle, etc. Your goal is to ask for the sale, the appointment, so you want to make note of exactly what you say during each call. Record it, listen to it and tweak it if necessary.

You want to make note of how you prepared for the call, what areas you are challenged in with regards to asking for the sale and what you are doing to perfect them in all aspects of your life.

5) Abolish your bad habits! Try this… If you have a bad habit of not asking for the appointment (sale) each time you communicate with a customer, decide that tomorrow you're going to ask for the sale at least once. Set yourself up for success by putting reminder notes everywhere and by wearing a colored elastic around your wrist. Tell yourself this elastic is to remind you to ask for the appointment and go for it!

Next, celebrate and document your successes in your journal. Set a new goal for tomorrow and go for two! Just keep repeating these steps and before you know it, you have a brand new habit.

> **Resource:** You can learn more about how to use your SSJ and PSJ by reading your *Chris Mullins' Nuggets® for Sales* print newsletter, or go to www.GreatBottomLine. com for more information.

Here's an excellent example of the Sales Mindset taking hold with one of my students. Your phone rings….

"I need to cancel my appointment."

"I'm sorry, we don't take cancellations."

Wow, imagine that!

Chris, I had a customer call to cancel today. I said, "I'm sorry to say you cannot cancel that appointment" (obviously, my tone of voice is critical here).

[SILENCE]

The customer proceeded to say they had another appointment and had to cancel this one. I said, "I especially put this time aside for your appointment with Jen and I don't have another opportunity for quite some time."

[PAUSE]

They kept it — yeah! I do this on a regular basis now and it works.

Linda

SALES LESSON

I really like this strategy. As with all new ideas, test it, experiment with it and see how it works in your business. Warning: Do this with a smile, but with more than just a hint of being concerned for the prospective customer that they won't be able to get back on the list anytime soon. Be sure they know you're disappointed and concerned. Also, let them know that the people (in this case Jen) are now being inconvenienced — they've set aside this time especially for them.

When they do show up, give them the Royal Red Carpet Treatment. Be sure that they feel like a king or queen and how relieved they are now for not canceling, after all. This way they would never again think of canceling an appointment with you! Here's how to *not* sell an appointment.

I recently visited a local acupuncturist for a 15-minute complimentary consult and I was 10 minutes late. The consult ended up being 30 minutes long. The acupuncturist should have stayed focused on what she wanted to accomplish in that visit, which was to stick to the 15 minutes (since I was late). She should have rescheduled, because now if I do become a customer I will not take any of her "rules" seriously, including being late. This consult time is set up specifically for that, so if a customer is late they should have to reschedule, otherwise it defeats the purpose.

During the presentation the acupuncturist doesn't sell me on her, the services, what acupuncture is, why it may work, etc. Basically, she is telling me that there's no guarantee. She did provide testimonials, and when I asked she gave me vague ideas as to the different types of problems she has successfully treated. Interestingly enough, I happen to know that this acupuncturist has a good reputation, but the problem is she personally doesn't know how to sell, so either she shouldn't do the presentation or needs to get some coaching in that area.

POWER NUGGET™

As discussed earlier, have one person focused on sales

In addition, when I asked if insurance covers it, she was uncomfortable with the question and said, "You'd have to check with your insurance." Next, I asked how much the treatment cost.

It was $80 for the first visit and $65 thereafter. How many visits would I need? Would I start with four? How long are the visits — an hour?

She said she has a Monday night clinic if I was interested in that if the pricing is a little too much. I personally wasn't thinking that I wouldn't pay that amount, but didn't say anything about it.

I asked what a clinic was. She replied that it was a group of patients in the same room, that's all. She said she didn't give the same type of service as she would privately. (It would have been far better for her to sell me on the entire experience from the very beginning, including her private patient sessions and everything that included. That way, when she offered the clinic as an option, the differences would be clearer.)

It was very bad that she brought up the clinic when she did and the way she did with no way for me to know what was different between them. In addition, because she didn't take the 15-minute consult seriously with me being late and then giving me 30 minutes, I couldn't take her seriously that she wouldn't be giving the same type of attention to me during a group setting as a private.

Emily's Question of the Month Which You Can Adapt to Your Own Business

While this is an example from a dental business, it is one hundred percent transferrable to your business and all *businesses.*

Hi Chris,

I have a question for you. We have started advertising in the *Alaska Airlines* magazine, and because of that we have started getting calls from people desperate to get rid of their dentures who live quite a distance from us.

I have had calls from Oregon and today one from Arkansas. How would you suggest we handle these calls? I have been using the exact same phone protocol for these people so far, but when I tell them that we should get them in for a complimentary consultation, they say that they need to have an idea of how much it is before they make the trip to Seattle to come for just a consultation.

Of course, I tell them that I am not able to offer an exact fee, and I offer to send them a packet of information for them to look over if they can't schedule a consult right now. Aside from mentioning that they should call when they know they will be in the Seattle area so we can set up a consult, do you have any other suggestions?

Emily

First, this is more about you putting up barriers in your own mind than about the distance. The word here is desperate! The pain here is desperate. (Knowing the specific pain for each caller is critically important because this is how you'll remind them that they made the right decision and coming into your office is the solution to their pain.)

Don't forget to get into the habit of simply answering only the questions prospective customers ask in all your conversations. Don't give more information than that specific question needs — then PAUSE.

It seems natural for us to give more, but the caller doesn't care, most of the other information we give doesn't make any sense to them, they really are calling for an appointment.

Without spending too much time, you've got to find out specifically why these folks are desperate. For example:

Mrs. Smith: "I'm desperate, I can't take it anymore!"

Emily: [with genuine urgency in your voice] I'm so glad you called me today, you made the right decision. Mrs. Smith, what would new dentures do for you?"

Mrs. Smith: "Oh my goodness, I'd be able to..."

Emily, what happens here is you record in your computer for this inquiry the exact vocabulary the caller is using. So, if she says... "I can't take it anymore," you say, "I can hear in your voice that you just can't take it anymore. Dr. Generic has instructed me to get you into our office right away."

Then you go into the "3-step scheduling" verbiage. When the caller asks about money, you say...

"Dr. Generic will be meeting you in person on xx date and he'll tell you exactly how he'll give you [here's where you insert what they say would be different — the exact vocabulary they use] new dentures."

Then, jump right back to the CLOSE, the 3-step scheduling and PAUSE.

The Power of the Pause

After offering an option to the prospective customer, PAUSE. Give them a chance to respond to what you're saying. Many times I find that the front desk person will suggest, for example, an appointment time, then pause for a brief moment and start making other suggestions before the caller has even had a chance to respond.

Usually, I'll find this in folks trying to make changes in their presentation skills. However, office managers and owners need to monitor this closely or you'll just be creating new bad habits.

Slow down the pace of talking. One of the ways you can do this is to stand up while trying to book an appointment, for example. Another is to put a 3 x 5 lime green index card in front of you that says SLOW DOWN.

You don't need to say so much, especially if you tend to go into much greater detail than is necessary with prospective and established customers. The goal should be to build a relationship early on: "What made you call today," etc. Then make the appointment, confirming for the caller that everything they just shared means that they made the right decision in calling your office.

Chris' BIG Thought

Be careful: Too much information (TMI) is the fastest way to guarantee talking yourself out of a sale, which in this case is closing the appointment. Many times, if we don't hear a YES right away from the prospect, we get nervous and start to talk or even downsell our original offer. You must be sure that you're not going to sell your appointment, product or service at that time before you go into a downsell, perhaps signing them up for your free customer newsletter.

AHA Notes:

Chapter *Three*

Scripts: Get Mullinsized

This is your cheat sheet. Review this sheet often; keep it next to your phone. Practice this outline during your Weekly Sales Drills™ together.

10-POINT SCRIPT OUTLINE CHEAT SHEET™

1) **Greeting.** "Thank you for calling XYZ business. This is Chris speaking. How may I help you?" [All done with a smile that says we're here for you.]
2) Get the **name of the caller**. Just ask.
3) **Use the name of the caller during the call:** *"Fantastic, Mrs. Smith, let me be the first to welcome you to XYZ."* Basically, you're assuming the appointment, the sale, the invitation to visit.
4) **What prompted you to call today** (identifying the problem/emotional pain)? Listen, document and share this information. *"Terrific, Mrs. Smith, you made the right decision in calling XYZ today."*
5) **Repeat the problem** to be sure you understand and they can be reassured that you get it. Repeat back what they said in their own words.
6) **Provide the solution — make the appointment**. Sales opportunity #1. [Refer to your "how to" schedule sheet.]
7) **Get contact information** — Sales opportunity #2, always at the end of the call, now that you've got the appointment and have established some trust. If, for some reason, you're not able to make the appointment, you still want to ask for this information so you can stay in front of them to remind them that you're here. Example: *"Mrs. Smith, since you're not ready to make an appointment today, I'd like to send you our free customer newsletter. Let me jot down your mailing address and I'll get last months' issue out to you right away."*

   Now you have their mailing address and if you stay in front of them via your mailings, post cards, birthday cards, etc., you'll get them in.
8) **How did you find us?** Listen, document and share this information.
9) **Close properly** — *"Thank you once again, Mrs. Smith, you made the right decision in calling us today. We look forward to seeing you tomorrow at 2:00."*

Resource: For an example on how to use scripts, go to www.UniversityOfPhoneProfessionals.com.

10) **Listen for opportunities** and share the information with the team.

Remember, you're always selling. Assume the sale (the appointment) by moving right into scheduling the appointment regardless of the reason they called. Your phone represents an appointment (a sale) each time it rings.

Resource: To learn about our script software tool, go to www.UniversityOfPhoneProfessionals.com.

Here's an example of how Nigel Worrall of Florida Leisure Vacation Homes uses their 10-point script outlines. Notice they have one even in the bathroom! From top down are Kim, Tiffany and Tracey. Check them out at www.florida leisure.com.

Did you know that the highest earning sales professionals focus on the exact language that they use to sell their services and prospects? They actually keep records of the vocabulary each of their clients and prospects use so that they can use the same language when speaking with them. This obviously helps your prospective customers to feel like you get it, you understand them, they won't be intimated, trust will begin and as long as your team members are on the phone with new customers to discuss their wants and needs, of course staying focused on closing the appointment, the more of a guarantee you have to booking a qualified appointment. In other words, not only will that customer make an appointment, they'll show up and if you deliver the same type of sales coaching I'm teaching here to the rest of your team members, you'll have an even greater opportunity to upsell to the same customer with great results.

You don't just want a team member who can book appointments because they say they're a people-person, you want individuals that are coachable, that understand your business, that in order to grow, you must have x number of new customers daily. You need individuals that can engage the new customer in conversation, get inside their head, understand specifically why that person called your office TODAY!

You've also got to teach your staff to not overstay their welcome, otherwise they'll unsell the already made sale. In each selling situation, there is a specific "sweet spot" where sales peak, stop short or go long and suffer. This is another good reason for tracking the details of all your leads, every single telephone call, e-mail, regular mail and fax.

Know Your Numbers

Let's say you have a staff of individuals that love people, have sold in the past successfully, even understand the technical aspects of the business. You still must make sure that the language they use over the telephone and in person is carefully crafted in a SCRIPT that gets memorized. You don't want your staff to read the script word for word, but you do want them to use it as a guide.

Sadly, few businesses, few professional services of any kind operate with polished sales scripts for even their most important functions or steps. Sales professionals are permitted way too much liberty in freelancing and adlibbing their presentations. Hardly anybody scripts — except, for example, ALL highly paid, highly successful trial lawyers, who script and rehearse closing arguments. Too many business owners insist their employees refuse to learn, role play and stick to scripts. Actually, the truth is they'll kick and scream mostly if you don't coach and train them on the purpose of the script. Many will just give it to them and say, "Good luck."

You must develop strategies to guarantee that your business will stand out amongst the rest. The telephone is the most important tool in your business, this is where you begin!

A very big part of the magic is that, from start to finish, expert staff never deviate from language treating each call as very real. No incongruity. No slip-ups. Not even a wink. And the language is as carefully crafted as if done for print by a top copywriter. For an example on how to use scripts, go to www.UniversityOfPhoneProfessionals.com.

There are 10 steps, 10 different areas that you need to think about, focus on and implement in your script and in your business. Everything that I'm going to discuss with you is important and you need to apply all of it, but when you're reading, sometimes, all of a sudden, you may hear something that sounds like a bell going off in your head. These are things that really kind of made you go, "Aha!" or excite you.

Make a list of those spots first and then after each session make a commitment to put a step in place — an action step — which achieves that list because those are the things that you'll probably get accomplished the quickest and have the best feelings about to get your confidence up when it comes to tackling anything else on the list.

In addition, give yourself a deadline. You can't just make an action list of things that you're going to fix. You've got to have what you're going to fix and how you're going to fix it, but you also need to put a deadline on it to help you commit to doing it. With that, let's begin with Getting the Greeting Right.

Step 1 — Getting the Greeting Right

Let's talk about the telephone greeting. I think that this is probably the biggest area that we're having a really hard time embracing. Embracing the importance of getting it right over and over again. As mentioned earlier, when you go to a play, you sit there, excited and ready to be entertained. Even though it's the last show, you expect to get 100% from all the actors, the same attention and enthusiasm and excitement and feelings as when they did the first couple of shows.

> **Resource:** For updates on this script, go to www.UniversityOfPhoneProfessionals.com and www.GreatBottomLine.com.

You don't want to hear from anybody, "Wow, these guys have been working every single night and this is the last night so it's understandable for them to be not doing quite as good." That's not OK to you. You scheduled time to go and it's important to you. It's really the same thing with the telephone greeting. There's a tone of voice that has a lot to do with it; your tone and your voice in general and the language that you use. You need to be consistent.

When I teach the greetings to owners and other types of clients and staff, I hear, "Chris, why do we have to do it that way? Why can't we do it a little shorter?" First off, I'm an expert — a telephone expert — and I'm The Phone Sales Doctor™. I've been doing this for 25 years now. I see what works and it's pretty consistent. The most successful businesses do it this way.

POWER NUGGET™

Treat your established customers like new customers

It is important that you focus on the greeting and the greeting needs to be inviting. If a caller is calling your business, the prospective customer (and even the established customer) feels that just by hearing your voice, your words and your tone that they're important to you, that you're waiting for them. That's the kind of feeling they need to have.

When the phone rings you want to say, "Thank you for calling ABC. This is Chris speaking. How may I help you?" Then you pause and you don't say another word. Everything you say, the words you use, the tone of voice, the rhythm in which you say it tells the person calling, "Wow, they want to talk to me. I'm not interrupting them." Even before they know anything about your business, they've decided that they've called the right place just by your greeting.

After they hear that kind of greeting it turns to a different frame of mind of what they're really doing and what's about to take place.

In other words, my point is that the person answering the phone decides that they're not comfortable so they don't do it that way. You have to decide that this is going to help you and the caller and you need to agree that this is how you're all going to answer the phone. You cannot have team members decide that they're not going to do it that way, that they're going to chop up the greeting, that they're going to do it differently because they might be uncomfortable or embarrassed. I say this because it happens.

The greeting needs to be consistent. To help the person answering the phone focus on that call, you've got to remember the average potential revenue that you can bring in with one new prospective customer. Let's say the average is $1,500. By the time the greeting is happening you're potentially losing that possible revenue.

You want to have a strong smile and the appropriate words, and you want to have your 10-point script outline in front of you at every possible location where somebody could pick up the phone and answer it.

The next step in the greeting is to **pause** after you say, "How may I help you?" You don't want to say anything else after that. The reason I'm emphasizing this is because you're busy and, in most cases, the person responsible for the phone has all kinds of other things they need to be doing. The greeting is also your transition phrase and that's one of the reasons for the pause.

POWER NUGGET™

Much of what I'm teaching can be used in outbound calling

The primary function of the person answering the phone is the telephone. They must be an expert at that and it must be their top priority. Everything else is secondary and you need to make sure that your business is set up that way. The phone is the most important position. The front desk expert has the most important position in your business.

That person needs the support and guidance of the owner. The same thing goes for people standing around that particular individual putting things on their desk while they're on the phone. Even that little moment is a distraction to them. Remember, this is a sales presentation. Have a system in place where there's no putting things down in front of them to read while they're talking on the phone, there's no sign language from other people standing around them, perhaps the owner or manager or team member trying to say something to them. They've got to focus on the phone and you need to have your business set up to support that person and help them focus on getting the greeting right.

After you pause, listen to what the caller has to say as it is very important. I'm emphasizing the pause because we're conditioned to hurry up and move on. We mentally jump into the next task on our list because we're trying to hurry the call along.

When you get the greeting right and remember to pause, you're helping yourself slow down. And you're helping yourself listen to what the person who's calling is about to say, so the greeting is for you just as much as it is for the caller. Each word from the start to the finish of the greeting is helping you to mentally shift, focus and move into the task of the moment — the caller.

POWER NUGGET™

Designate a phone person

Now you've listened to what they have to say and you know you're not talking over the caller. Usually we don't pause, we kind of go on to the next thing in our conversation, sounding like we're on automatic pilot and talking over them.

When you say, "Thank you for calling ABC. This is Chris speaking. How may I help you?" you're waiting and you're listening. By doing it this way, you're relaxed, you're not rushing anymore and you're feeling much more comfortable. This makes the caller feel exceptional. They're not expecting it.

What you're going to do next is document quickly, word for word, in their language, what they just said. "I'm calling about a [service]," or "I'm calling to make an appointment," or "What are your hours?" Pause, listen to what they say, and at the same time quickly type what they say using their language and their vocabulary.

Every time the phone rings, you're in a state of readiness, ready for that caller. You have your new customer information open or your record open on your computer, and you're ready to not only use your guide, not only use your 10-point script outline, but you're ready to immediately type what the person is saying to you.

This is about standing out amongst the rest. This is about being the diamond in the bunch where you're being compared to everybody else. You're not just being compared to other owners, you're being compared to every single business out there the caller has just done business with.

You're going to break the cycle because you're going to dazzle them from the time they call you with your greeting, your words, your tone, your rhythm, your pausing and your listening.

Now you're going to ask for the appointment on this call. They're going to go from the due diligence of asking questions to getting things taken care of for themselves.

That's really the point I'm trying to make with you here — you want to be that diamond in the bunch. You want that caller to feel like, "Wow, that was an incredible experience." You want them to talk about you at the dinner table at night. I think you'll find that when you use the proper greeting, you're being consistent, you're pausing, you're not hurrying them along and you're not talking over them, they're shocked. They're not expecting it. It tells the caller that you really paid attention.

I know that sometimes you might be thinking these 10 points are going to take a really long time. It doesn't. The whole idea is you want to have your new customer phone slip or your script or outline in front of you ready to go. What you're probably going to end up doing is take the 10 points here and edit them a little bit. You want to practice it and read it a few times a day.

Another thing you want to be careful of is speed, so you don't want to talk too fast. You want to do it slowly and you don't want to be doing it eating lunch or chewing gum at the desk.

A lot of times people finally get the greeting down pat but they're not thinking clearly on the real purpose of the greeting which is a tool for yourself and a trust mechanism for building a relationship with your established and prospective customers. Everything I'm teaching can be used for all of your customers and it should be systemized and automatic. The more systemized and automatic (and consistent) you make things, the more you're running a business where that's the expectation and the base foundation for everything that you do.

When you start a new program or project, the question in everybody's mind at the business is, "Well, what's the system?" because it's their guide, it's their safety net. It keeps everybody running efficiently and it runs your business better. You've perfected it to make it be what works.

You also want to remember the bottom line is this is a business and my belief is sales is customer service and customer service is sales. You can still love what you do, you can still love your career, you can still love to be of service to your customers and to be in the business and still believe that the bottom line is important everyday. What did we sell, how many new customer appointments, how many showed up, how many followed through? It is about business and we've got to understand that from the front office to the back office.

So, the telephone is the tool; it's the critical tool, it's the most important position and needs to be treated that way. Remember, if your average potential sale for one new customer in your business is $1,500 and we're not doing this right and systemizing it, then we're basically saying it's ok if we miss it, it's ok if we lose the prospective customer, it's ok if we lose the $1,500. That wouldn't make a whole lot of sense.

The other thing I want to say is you've got to be careful about the folks answering the phones, because if they don't speak clearly (for whatever reason — they might have some sort of speech impairment or maybe they have a strong accent or whatever that's not their fault) it doesn't help their position. It's hard for them to become an expert at their position if there's something going on with how they speak. When you call in, can you understand that person?

In summary, you want to get the greeting right, and getting the greeting right is pausing, listening to what they say, repeating quickly back what they say in their words, and being careful of your tone and speed.

Resource: One quick thing I want to mention to you is about headsets; I strongly suggest you use them — they will help you to do an exceptional job. It will help with fatigue and when you're trying to multitask. What I suggest you do is contact a vendor (or several) for headsets. (Go to www.GreatBottomLine.com to find out more about headsets.) It's important how they service you if something goes wrong with them, so you want to see how they treat you.

Step 2 — Getting the Name of the Caller

"Smile! Say cheese to increase sales (appointments) and performance."

A small, aggressive commodities brokerage firm trains its brokers in professional voice production, the same training that singers and actors get. In the sales pit everyday, they're screaming and yelling and could easily lose their voices. It's the equivalent of an account executive's computer going down. Smiling is also an essential skill for a sales person.

The sales director for one of the America's largest corporations hired JB Iden, a director of the New Stage to teach his salesmen to smile. Most of them thought they knew how to smile, but Mr. Iden convinced them that in many cases their smiles were merely smirks. He taught them that smirking involved only the lips. The entire face, and especially the eyes, is needed to accomplish a sincere, friendly and attractive smile. After Iden's smile clinic, the salesmen increased their sales 15% within three months.

Source: unknown

Imagine that! So that's something you want to do in your business on a regular basis.

My challenge and my big question to you is, "How far will you go?" Have your own smile clinics to remind yourself that everything is about behaviors and habits. You can't accomplish something without changing behaviors and habits and without measuring it — what gets measured gets done, right? So imagine if you had your own smile clinics at your business and your weekly staff or sales drill meeting, reminding each other about smiling, how much easier it would be to perfect the mindset and attitude that we're talking about in the 10-point script outline.

POWER NUGGET™

Imagine if you had your own smile clinics at your business

Let's continue with step 2, getting the name of the caller. Believe it or not, getting the name of the caller is a particular area that I find I have great difficulty getting folks to do initially. And the reason is they feel like they're being pushy. The front desk person, for example, feels like they are being pushy or nosy if they ask the person's name. They will say something like, "Hey, Chris, if they wanted me to have their name they would have told me." That's really more to do with how you feel about asking them their name. It's not about what they think, it's about what you think.

Here is something I learned from my mentor, Dan Kennedy. Write this down on a card: "You're not your customer." Get a 3 x 5 lime green psychological trigger card and write on that card with a fat black marker, "You're not your customer" or "I'm not my customer." Just because you feel uncomfortable and it might seem pushy or nosy to ask their name, doesn't mean they feel that way. In most cases it's not an issue.

When you ask for the person's name, you're controlling the conversation, you're continuing from where you left off with the greeting. It's important to control the conversation, because if you don't the call will go all over the place. This helps to build the relationship and it continues the trust process. If you are feeling uncomfortable, just practice it a few times. You will get used to it.

So basically you'll say, "Thank you for calling ABC's office. This is Chris speaking. How may I help you?" Let's say the person says, "I'm calling to find out if you do free estimates." You pause, listen and then you say, "I'd be happy to help you with that. May I ask your name?" "Yes. My name is Lisa."

POWER NUGGET™

What's important is your tone of voice

What's important is the tone of voice in the way you do it. "May I ask your name?" It's sort of like you're kneeling or bending down, looking at them, or you've got your hand on their shoulder, that sort of thing.

It's important for you to practice. That might not sound like a big deal but it is. If you follow each step and you do it with the right tone of voice, you'll be very comfortable. Get their name because you want to control the call. You're trying to quickly build a relationship. If you ask the name and you do it in the right tone of voice, it will not become a problem.

You know, it's tough out there for consumers to trust businesses. We tend to be afraid of salespeople whether we call them over the telephone or drive into a car dealership and find everybody is all over you.

You have to do everything you can to get it right when they call, because if they hang up and didn't make an appointment, they're gone and you will never see them. Getting the greeting right and being comfortable with asking for their name is critically important.

Resource: To learn about our Certified Phone Success Advisor classes, including the 10-Point Script Outline Cheat Sheet, go to www.GreatBottomLine.com.

Step 3 — Using the Name of the Caller

Number 3 of the 10-point script outline cheat sheet is "Use the Name of the Caller" during the call. This is probably some of the most important stuff that we're going to talk about. We've done the greeting and we've gotten the name of the caller and now we're using the name of the caller.

You're going to do this throughout the call. It's not to be missed, so please focus carefully on it and notice what I say. Here's a complete example on using the name of the caller.

"Thank you for calling ABC. This Chris speaking. How may I help you?"

"Are you taking new customers?"

"I'll be happy to help you. Did you ask if we're taking new customers? May I ask your name?"

"This is Mary."

"Fantastic, Mary, let me be the first to welcome you to ABC."

If you're not comfortable saying fantastic, you can say terrific or something to that effect. You do want to say a positive, exciting word. I usually teach clients to say, "Fantastic Mary, let me be the first to welcome you to ABC." I use her name and am very enthusiastic and excited. But I also have not just the right tone of voice, but the right rhythm and right speed. I stay consistent throughout the whole process and then what do I do? I pause for a moment. I just finish my sentence and stop.

Here's what's important about number 3. For the most part the prospective customer calling is not always ready to make an appointment. They usually say, "Oh, I'll finally get around to calling the owner and find out about making a new appointment and what's involved in that." They feel that they're going to go through the list of questions, but the point is deep down they want to make an appointment. That's why you're going to be in the assumptive mode.

So when you say let me be the first to welcome you to ABC's office or business, you're saying automatically that they're in, that you are accepting them, that they're welcome, that we're going to make the appointment. But that's not what you want. You need to be on automatic pilot. Remember, your business has spent time and money, lots of money, on marketing to get the phone to ring with the right prospective customer.

Now we've got to make sure that the person on the phone, the front desk expert, is ready and armed to recruit that customer on the spot. We need to turn that call and control it in a positive way.

Let me be the first to welcome you. In the mind of the caller, it's like, ok, maybe I'm making the appointment now. In addition to that, it continues to keep the front desk person focused and armed with what you need to be doing, one call at a time, and that's scheduling appointments correctly. Just because we're doing all these things right, if we miss a beat, it doesn't mean they're going to show up. This 10-point system is designed to also help with the show-up process. That's what you really want to think about here.

I don't just want to give you ideas of ways to use the telephone without telling you why. This continues with the flow of the call, which is to get them to make an appointment, to get the decision made for them, to dazzle them, to stand out amongst the rest, for them to feel like, "Wow, I've never had an experience like this before." How many businesses do you know that answer the phone the way we've talked about consistently, automatically, and how many do you know that say, "Let me be the first to welcome you to ABC"? In the right tone and the right rhythm, never mind the words? It's critically important.

So, we're trying to change the entire process. And that is exactly what you're doing with number 3. And you know, there's no question that if you're going to do business, improve the bottom line in your business, add even more customers, you want to follow this format. "Thank you for calling ABC.This is Chris speaking. How may I help you?"

"I'm calling to find out about making an appointment."

"Ok, I'll be happy to help you with that. May I ask your name?"

"This is Mary."

"Fantastic Mary, let me be the first to welcome you to ABC."

See, you're controlling the call, you're monitoring the flow of the call and they're riding along with you because of how you're doing it. And what you're also doing that's really important to note here is you're telling them, in so many words, by using their name, with this phrase and beginning that process early on, "You made the right decision, Mary, in calling us today." One of the things you want to know is that a lot of people have a difficult time making decisions. They're concerned about making the right decision and they go back and forth and procrastinate wondering if they're making the right decision. And because of that they can take a long time to get around to making any decision. When they finally call you (they've made the decision to call), you've got to pull them in right then and there and get all of these points right.

What you have to do is be an expert at every step and focus on practicing every single step. Keep in mind that you want to have your 10-point script with you. You want to keep it nearby and you want to practice it and look at it on a regular basis.

Step 4 — What Prompted You to Call Today?

Tilt your head ever so gently, use the tone of your voice to work for you and picture what the caller looks like.

Step 4 is very important; this is where the caller gets to tell you their emotional (not physical) pain and why they felt it was important to make the call today. Identifying the problem and then reassuring them that "they made the right decision" is very important because now you're turning this call into being about them, building a relationship and establishing some trust. Even asking questions like this will show the prospective customer that you're different, that you care enough to ask.

Listen to what they say and document it. Later in the conversation you'll want to repeat what they said in their own words and if they don't schedule right away, remind them in their own words why they called today.

Step 5 — Repeat the Problem

It is important to remind you of PAS — Problem, Agitate, Solution. Number 5, repeat the problem, is when we start to agitate and I'll explain that to you in a moment.

Before I do, read this quote because change can be difficult and challenging for folks. As a teacher, I'm faced with that on a regular basis.

Creating new behaviors and eliminating bad habits can be difficult. You need to have an open mind because we're in business to stay in business and to grow, and unless we stay in business and grow, we cannot help our customers.

Here's that quote. It's titled Attitude by Charles Swindle.

"The longer I live, the more I realize the impact of attitude on my life. Attitude to me is more important than education, than money, than circumstances, than failures and what people say or do. It is more important than appearance, giftedness or skill. It will make or break a company, a church, a home. The remarkable thing is, we have a choice, everyday, regarding the attitude we embrace for that day. We cannot change our past. We cannot change the fact that people act in a certain way. We cannot change the inevitable. The only thing we can do is play on the string we have and thus our attitude. I am convinced that life is 10 percent what happens to me and 90 percent how I react to it. And so it is with you. We are in charge of our attitude."

(For a copy of this quote, go to www.GreatBottomLine.com.)

Your prospective customer on the phone needs to be reassured that you get it. They're opening up to you, so it's really important that they continue to feel as comfortable as they have in previous steps.

In repeating the problem, you need to stay focused and you need to keep the flow of the conversation going, because you can get off track. At the same time you want to be sure that they know you got it because this is where they are really being open. So you're repeating it, you're agitating it.

What do I mean by agitate with Problem, Agitate, Solution? In step 4 you're finding out what the real problem is as much as they're able to say it to you, but you're getting more than just them calling and saying they called for an appointment. You're agitating it, because you repeat it back to them in their words. So you're going to have a sense of urgency.

When they call, it might not seem urgent to them. They know it's important, but it's not really going to seem urgent and they may be just getting around to it. They don't really understand the details of it, but if you make it not seem like the same kind of calls that you get day in and day out, if you make it seem really urgent and critically important, then they're going to kind of sit up straight a little bit more, and they're going to be like, "I guess this really is important."

You want to listen and repeat the problem they were calling for, in their words, with some urgency and concern. What you're really doing is beginning the agitation process. If you don't do that it's just not going to be as important to them and they're not going to get around to making the appointment or they might not show up for the appointment. This is really about the show-up rate.

You want to continue sharing this information with everybody else in your business that would potentially touch a customer — talk about it.

I'm going to try to play this out for you from the top:

"Thank you for calling ABC. This is Chris speaking. How may I help you?"

"I'm calling about making an appointment possibly, or maybe you could tell me if you're taking new customers."

"Absolutely. I'd be happy to help you. May I ask your name?"

"My name is Mary."

"Terrific Mary. I want to be the first to welcome you to ABC."

"Okay."

"Mary, may I ask you how you found us?"

"The Yellow Pages."

"Okay. Mary, what prompted you to call today?"

"Well, to tell you the truth, my husband has been really concerned. We're not really excited about looking for new insurance. I have a problem with my current plan. I'm paying too much."

"Okay Mary. You're saying that you and your husband… what's your husband's name?"

"Jim."

"Okay. You say that you and your husband Jim have been really concerned about how much you're paying for your current insurance. Is that right?"

"Yes."

Hopefully you see how I did that by going through the whole flow. You want to change the tone of your voice, more in an agitating mode, more of a concern, really highlighting an urgency with this particular call. You really want to perfect that. Over time and with business, you'll get much better. [Go to www.GreatBottomLine.com to get info about our online training and to hear Chris' demo.]

One of the things I recommend that you do is record your calls. Check with your state to find out what the guidelines are for recording calls for training and quality assurance. Each state is different.

The idea is for the front desk expert to record and critique their calls to sharpen their axe and perfect their craft. Have fun with this; compete with yourself. Maybe just a couple of days a week, or even one day a week, rotate the days to see if you sound different on different days and record them at different times of the day to see if you sound different. Record them and then play them in the weekly sales drill meetings. Play them back with the team so you can hear them. That way you can hear how you're doing with your voice and your tone and the inflection in your voice. And your teammates can critique you.

Many of my clients do this, and they find when they do all the things I suggest they have the best results.

Again, step 5 is only the first part of starting to agitate the process and using it. Remember, you want to repeat the problem as much as you can in their words. You're beginning to agitate it, you're changing the voice and you're changing your tone. You're doing it with a little bit more of a sense of urgency. Don't forget to continue to use the name of the caller.

Step 6 — Provide the Solution

Step number 6 is to provide the solution — make the appointment. Here's what I want to emphasize about this more than anything. What you really have to remember and focus on in this entire process is that the telephone is a tool. It's an automatic tool and it's a sales tool.

Hopefully by step 6 you're able to see how important it is for you to follow the steps so that you can get the appointments instead of letting people just say, "Okay, I'll call you back." By doing that, you're letting them go thinking you did a great job because you were nice and pleasant on the phone. Hopefully you're seeing that it's about hanging onto that person and inviting them in to everything that we've said. Now you're going to make the appointment, which is the solution.

What I'd like to see you do when the phone rings is view it as, "It's my job to focus on scheduling the appointment and following the phone script outline. It's not my job to analyze why they're calling. And when I hang up, if I've scheduled an appointment versus them saying they'll call back, I'll know I did my job."

So provide the solution and make the appointment. Here is how you're going to do that. You've got to understand you have to *ask* for the appointment, and if you handle the call right, you're providing a solution and they're going to see it that way.

What you really want to be doing is provide a service for folks, but you've got to do it in a certain way on the telephone or they'll never get the benefit of being with the owner or the team.

Always have the first two to three available appointments in front of you throughout the day so that you don't have to spend time looking.

You want to be ready for each call so you're not talking too long and you're not pausing. It keeps you more efficient. Your timing is crucial. Another thing you want to keep in mind is you want to get the prospective customer in as soon as possible. If you have an opening today that's appropriate, put them in it. If you don't capture them right away, the more time you have in between appointments or from the time they talk to you on the phone to the time they show up for an appointment, the more apt they are to cancel the appointment, not show or change their minds. You don't want that.

POWER NUGGET™

It's not your job to analyze

What you want to think about with scheduling is speed. Getting them in quickly is top priority. You want to have all this organized and comfortable, to move down the process of scheduling an appointment smoothly and quickly. You don't want to take a long time looking around, it's unprofessional. The person on the phone doesn't want to take the time to wait for it. Scheduling is so important we have a script we use in my Phone Success Program.

Number 5 is repeating the problem and you're agitating it. Number 6 is providing a solution, you're making the appointment. And if Mary has a problem with the appointment and can't make it right then and there, you're agitating it a little bit more, "Well Mary, let me just make sure… didn't you say that you and Jim are really concerned?"

"Yes."

"And you have some concern about the price you currently pay?"

"Well, yes."

"As I said, ABC specializes in customers that are dealing with your exact concern."

"Mary, we really need to get you in here. I'll meet you, too, but I need you to come on in here as soon as possible. The next available appointment I have for you is today at 10 o'clock. Come on in and meet with us at ABC." You pause and you're quiet and you have some urgency and some concern for Mary in your voice.

You pause again to wait and see what happens and deal with the next step from there. You always want to do the soonest available appointment. Each time that you ask a question, be very careful. You'll notice through this whole program I've been talking to you about when you ask your question, be quiet and listen to the answer. You've got to train and condition yourself to do that.

A lot of this you can use even in your personal life and with communication in general. The psychology of it is really the same. It's something that you can do outside of the business with other individuals. It will help you on that level, but it'll also help you sharpen your axe even more.

So in step 6, you've provided the solution, which is asking for the appointment and it's continuing to agitate the process a little bit. You can see how getting the information and repeating it is really important.

Step 7 — Getting the Contact Information

Your first objective, the most important crucial thing to be done quickly and effectively, is to ask for the appointment and get it scheduled. That's what the whole call is about. When the appointment is scheduled with the customer, that's the deal, without that you have nothing. Then you can ask them questions about contact information.

When you're making the appointment in step 6, you're getting into the contact information. What I mean by that is you're getting all the necessary information at this point to secure the appointment. Or, the appointment is already secured and you're getting the necessary information that your business needs in case you need to call them back.

Let's say they don't make an appointment. If they don't make an appointment, you want to get the contact information because you've spent a lot of time and money and effort to get the phone to ring with the right kind of prospective customers. If they don't make an appointment, you've got to make sure that you have their information so you can stay in front of them.

What I like to recommend is, "OK, Mary, since you're not ready to make the appointment today, I'd like to send out our free customer newsletter. This is a free newsletter we're going to send to you. What is your mailing address?"

That's what you want to do. Always make sure you get the details about the customer; their history, address and the phone number at the end of the call after you've established a relationship. If you've got it included early on in your phone conversation, you don't need to do that, you've got to secure an appointment with them, build the relationship that gets you to secure the appointment, in that order, then you can ask them for that personal information.

For the customers that are ready to make an appointment, after you get the information you need from them, you still want to say, "I also want to let you know, Mary, that now that you're a customer of ABC, you're going to be receiving our free customer newsletter. It comes out every month and there are a lot of great things in there." You want to include that whether they made the appointment or not.

The longer you are on the phone, the less likely you are to book the appointment

So, you've secured the appointment and now it's time for you to get the information necessary. Examine the questions that you ask; are they all really important for the first appointment? You and I know that every business is different, so take what I'm saying and translate it into your particular business.

Notice what I did here. You're watching the tone of your voice, you're being friendly, you're treating her like she is a customer, even though she's not yet. Treating your prospective customers like customers and giving them the same experience right away will get them in faster and keep them customers for life.

The more you can stay in front them, the better for you because you're reminding them you're there no matter what.

I'm giving you a lot of different ideas here, not just the telephone. Number 7 is important; no matter what, you cannot hang up the phone without their contact information.

Step 8 — How Did You Find Us?

This is a very important question and here's how you're going to ask it. Don't forget that tone is very important; your tone of voice, how you say something, the rhythm in which you say it, and then pausing. "How did you find us, Mrs. Smith?" Pause. The reason why the pause is important is because you want to give the caller a chance to respond.

A lot of times we're rushing around, we've got all kinds of things going on and we haven't quite yet gotten all the new behaviors and habits down. They say it takes 21 days to form a new habit. We know we're supposed to say, "How did you find us, Mrs. Smith?" and we might end up jumping ahead.

One of the most important things for the front desk person is for them to be able to focus and concentrate on one call at a time, one prospective customer at a time. When that call comes in it's really important to be able to focus on it, because we want to get that new customer in.

We're in this business because we want to be of service and help people. We want to solve their problems but we've also got to remember the business part of it. (You've got the marketing down, you figured out how to get your phone to ring with the right type of prospective customers with the right caller.)

Now we've got to deal with the phone so we bring them into the business. As much as we want to be of service, you're never going to get to be able to shine and deliver those services if they don't come in the door.

So your desk and your telephone are your business. Because they're your business, it's important that at the end of every day you go home and say not only was I a top-performing person but I did everything that I had to do that was absolutely necessary to focus on each call at a time. Bear with me for a minute if you're cringing right now saying, what is she, crazy?

POWER NUGGET™

Front Desk Experts – Don't allow the owner or other team members to distract you from focusing on one call at a time

I understand what it's like to think that you're doing one thing and find out it's something else. I understand what it's like to have three, four, five people standing around you and juggling all kinds of things and getting different messages. It's very stressful and hectic. But the bottom line is you are the front desk expert. This is *your* business. What I'm trying to coach you and teach you to do is to own it.

By that I mean provide solutions. If you find that there are people hovering over your desk and you're having a hard time giving your full attention to the calls coming in, you've got to make note of it and come up with a solution. Maybe it's your own time management and efficiency skills. Right now it's okay for you to be frustrated that you can't do what you want to do for each customer that calls, or each customer standing in front of you, therefore, you should get excited enough about it that you want to come up with solutions.

You have to give one hundred percent to your 10-point script outline, to practicing and to getting it right. You can do it, and you'll be excited about it. "How did you find us, Mrs. Smith?" If she says, "I saw you in the Yellow Pages, or a friend gave me your name on a post-it note," you need to write it in her language and her words because you'll repeat that back to her and that's going to be important to her.

"How did you find us, Mrs. Smith?" is only important to you to know because you're trying to maintain the position of building a relationship and building trust and reminding her that she called the right place, you care enough to ask, which means you're standing out amongst the rest.

Hopefully you folks are doing weekly sales drills. The reason I call them weekly sales drills is because I'm trying to help you build a culture that you are really in the sales business. We have total integrity. We are being of service to people with total integrity and honesty, but the bottom line is nothing happens until something gets sold, so we have to sell.

POWER NUGGET™

Sales Are Made Everyday

Someone will close — you or them — what's it going to be?

Whatever the reason they're calling you, you're supposed to influence them to come in. That's what you are supposed to do, and if you don't do it, you're out of a job. You'll be out of work, and the business will be out of business. You've got to really think of it that way.

So, in your weekly meeting you're talking about each customer that calls, and the prospective customers that are coming in for their complimentary 15-minute consultation (if that's what you do) — whatever window of opportunity you created for them to come in. It's important to share that information with whomever's doing the marketing at your business. It's important for everybody to know that these are how customers are coming in to see you. Here's why they're coming, what they're saying. Here's how they found you, because that can help your marketing. It can help you determine whether this part of the marketing is really working, or this part isn't. So it's important to ask for that information as well. Ask the question, then pause, listen and document that information.

Resource: To hear the audio interview with Chris Mullins and Ron Sheetz (the Done For You Video Marketing coach) and to get Ron's free offer, go to www.GreatBottomLine.com/marketingvideocoach.

Step 9 — Close Properly

Step number 9, Close Properly, is probably the most important step to help with people showing up for their appointments. As I said in step 7, you've never really stopped selling yourself or your business. I want to reiterate to you that there's no guarantee your prospective customer is going to show up just because everything went smoothly. And you've got to do everything you can from the beginning to the end of the phone call, as much as you can possibly influence, to make that happen.

Step 9 is an important step to help you do that. Even if they do show up, it doesn't mean that they're going to continue showing up and continue that process, so you've got to keep that in mind.

This is the end of your call when you say, "Ok, Mary, you made the right decision in calling us today and we look forward to seeing you tomorrow at 2:00.

Then say, "Mary, this time has been reserved just for you to meet with Mr. X." Pause, be quiet and wait to see what she says.

"Ok, Mary, thank you very much, we'll see you tomorrow."

And that's how you close. Now there are some important items we just mentioned here with the closing. You're using Mary's name, you're thanking her again, you're reminding her at critical points of the conversation, and remember that the prospective customer can be any type, it doesn't have to be someone that says that they're afraid; you would still follow the same program.

What are you doing there? You're reminding them of the time and you could also insert here, "Do you have a pen handy that you're writing this down with, Mary?"

"No, I don't, I can remember it."

"Well, I can wait, if you want."

You're trying to do everything you can to make sure this happens and you do have some influence. You can't be afraid or uncomfortable. You've got to work on doing extra steps like asking Mary if she has a pen handy. We're trying to get the customer to make the appointment, secure the appointment, show up for the appointment so we can be of service. And if you do it with the right tone of voice and the right rhythm, you'll be fine. It'll seem natural to you and for Mary, too.

POWER NUGGET™

"You made the right decision"

Resource: We do have two secret script closing weapons that we didn't include here. Go to www.GreatBottomLine.com to learn how you can get access.

Step 10 — Listen for Opportunity and Share

Here we are at step 10 — listen for opportunity and share the information with the team. During a conversation a lot of different things come up. Maybe somebody just moved into the area, maybe they need to come in for an oil change, but they also think they need to have a tune-up. Maybe they're saying that they've been really busy because they're getting ready for a special occasion.

Focus on the flow of the call and securing the appointment for the initial reasons the person called. Stay on that and document what they say.

You want to document as you have been right along and share it with the appropriate people on the team, especially in your weekly sales drill.

I'm really trying to coach and train you to think of the fact that you really are always selling, and selling isn't the "s" word. A lot of us think that because of the negative stigma attached to salespeople. As I mentioned earlier, when you really think about it, nothing ever really happens in this world until somebody sells something. First and foremost — sell the appointment.

First getting your marketing right, getting the phone to ring, and when the phone rings, selling the appointment. We're talking about sales as staying in business, as growing the business and being of service to all customers, and to do that we have got to stay thinking like a business. We have to sell, and it begins with the telephone.

Each time the phone rings, it represents an appointment, a sale. I know a lot of times the front desk expert is an expert in other areas of the business. It's very common but you need to be able to shake that off when you're in scheduling-the-appointment mode. Here's why. If you don't shake it off, you'll tend to diagnose them and you'll be saying a lot more than you need to say in the call. So, shut that knowledge off and just schedule the appointment. With the new customer example, talking about price is not your focus.

If somebody says to you, "How much is it?" you would say to them, "That's a good question and that's a question for Mr. X to answer. He's going to talk to you about that when you see him in person." All businesses have the same challenge.

Even when you have new people, they can focus on a few different steps instead of the whole thing at one time. I'm going to end this number 10 step with one more fantastic quote.

"Weak is he who permits his thoughts to control his actions. Strong is he who forces his actions to control his thoughts. Each day when I awaken I will follow this plan of battle before the forces of sadness, self-pity and failure capture me. If I feel depressed, I will sing. If I feel sad, I will laugh. If I feel ill, I

will double my labor. If I feel fear, I will plunge ahead. If I feel inferior, I will wear new garments. If I feel uncertain, I will raise my voice. If I feel poverty, I will think of wealth to come. If I feel incompetent, I will remember past successes. If I feel insignificant, I will remember my goals. Today, I will become a master of my emotions."

And that, my friends, is by Og Mandino, the greatest salesman in the world.

Resource: "The 10-point script outline cheat sheet is absolutely priceless and when used exactly as illustrated, it can't help but add additional volume to any business office." – Bill Glazer, President, Glazer-Kennedy Insider's Circle, www.dankennedy.com

AHA Notes:

Chapter *Four*

Business Owners Fixing Their Own Heads

"Would your team follow you into any battle? If not, why? If yes, why?"
– Chris Mullins

Why You Need a Coach, Trainer or Mentor

More often than not, regardless of experience, we're too close to what we're doing to realize that there are diamonds in our own backyard.

This is why having a great mentor, coach or trainer is so important, even for the big shots who are already making hundreds of thousands to millions of dollars per year. Having a team of coaches to look over your shoulder is the fastest way to even more growth than you're already experiencing and keeping your own individual axe sharp.

When HL Hunt (the oil billionaire) was once interviewed, he was asked the secret of success. Here is his answer:

1) Decide what you want to do
2) Decide what you'll give up to do it
3) Decide your priorities
4) Decide to do it

Warning! Don't Let Your Team Hold You Hostage

Stop making excuses for your team! Stop being lazy about getting this right. Truthfully, it starts with you, you must embrace the phone and decide once and for all that you will hold your team accountable for using scripts and closing appointments. If you do, you'll get incredible results — fast.

Bottom line: You're the owner, the business owner, and you run the show. So, just in case from time to time complacency has shown its ugly head, you can use this as a reminder.

As the business owner you must pay attention to the following information. The big question is, are you tracking your inbound phone calls?

Do You Know? Does Your Staff Know?

How many calls you get daily? How many appointment opportunities that didn't get sold daily? Why they didn't get sold? How many appointments were sold? How many prospective customers showed up for the appointment?

Do you know what happened during the in-person sales presentation for those that did show up? What happened once they crossed the threshold? Did your prospects and customers leave with a new appointment and plan for the next meeting, or did they tell you they'll get back to you and you said, "Okay"?

When was the last time you called your own business to play prospective customer? Then, after you fell off your chair, did you make notes of what you wanted to see done differently? Did you share that with your team in a productive sales drill or did you just sort of mention it and that was it? What happened to the list of prospects that didn't show up?

Be Afraid, Be Very Afraid!

Excuses, excuses, everyone's got one and they all stink! It's shocking (even frightening) the number of businesses out there that agree to dig deep to find out how their phones are being answered and when they find out the shocking news, they say... "Oh, it's not always like that" or "They're new" or "I know it's not right but we're so busy I'm going to let it go because if they quit, or get mad, it will be worse" or "We're understaffed" or "It's vacation time." One excuse after another!

You must inspect what you expect all the time. Even for those of you that are members of my Phone Success, Phone Sales and Mystery Checkup Call programs, as the owner you should mystery call your office at least once each week. It will keep you on top of the pain and successes and it will keep your staff on top of it because not only do you have our checkup call team looking over your shoulder, they know that the owner is doing it, too. In fact, you could assign members of the team to do weekly checkup calls and report the findings to you. This would be a great way to get support from the team because now they're part of the solution. In addition, your team knows you're involved. It shows them that you support it and you believe in their ability to continually improve.

Here's what's really happening in businesses that are not doing mystery checkup calls to provide support and coaching for the team on how to get it right call after call. I know some of you have been working hard at it, but be careful, don't get complacent! This little scare will be good for all of you. The person answering the phone...

1) Despises the "S" word (sales) and the "T" word (telemarketing, i.e., outbound calls); doesn't want anything at all to do with it. "Hey, they know where we are, they'll buy when they're ready. I'm not the kind of person to push them."

2) Believes that the owner is stealing from customers, charging too much, being dishonest, very unfair and taking advantage of people by selling anything at all. They put their own personal thoughts and feelings onto the client and prospect.

3) Price Threshold — Have financial challenges themselves and believe without a doubt that their prospects and clients have money problems. They can't afford the services or products and should spend their money better in other ways. Here's an example: My morning ritual is to ride my bike, then go to the local coffee shop with my 3 x 5 lime green card to write my intentions out for the day. The other day I ordered my shots to be added into my coffee and the cashier said to me, "Expensive habit." This is a perfect way to stop the customer from visiting your business. Would the owner know this was happening? No, not at all, but this sort of thing happens every day and that's a lot of lost sales opportunities.

4) This one is huge: In fact, read my personal story on page 6 or go to www.Great BottomLine.com to learn how I used to feel the same way. The person answering the phone feels it's their responsibility, their duty, to save the customer money. One way this is encouraged is to tell the customer to go elsewhere, "You can get it cheaper down the road." Every single one of our clients, not just owners, has this happen in their business. I know because I personally hear the recordings. Don't think for a minute that "Mary would never do that, she's been with us for 10 years!"

5) Resents the owner's wealth. Why is *he* driving that Jaguar?

6) RED FLAG: Even if just one employee thinks this way about your business — whether they answer the phone or not — it's a recipe for putting you out of business.

7) Sabotage all internal and external training and coaching efforts. "It doesn't work that way here," "We don't do it like that," "We tried, it doesn't work" or "We've never done it that way!"

Your team is complaining about the training they have to do (the mystery calling, the recordings) because they know you'll fold, you'll give up, you'll cancel the program and go back to the way things were. No monitoring, no measuring, no Sales Talk™, no sales drills and a major hole in your bucket.

An Observation

Chris,

Every situation is a sales situation, every moment is a sales moment. This is especially true in a dental office, whether you want the customer to book an appointment for the crowns they require or you're doing PR to create good will. Every act of kindness on our part, joke cracked or compliment paid the customer is like money in the bank against the day when something could happen like a suture untying or temporary arch of crown coming off.

Hopefully these things don't happen very often, but when they do you want to have earned enough brownie points with the customer that they are OK with it, rather than having them walk out of your office angry. Good communication is key. It may only take a minute but it is essential.

Dr. Sajner

How can you translate the above lesson into your own business? Dan Kennedy says, "Courage comes from deal flow. Deal flow is king. Create and attract a lot more demand for you and your services and products than there is supply or than you need to achieve your wealth objectives and you will naturally become emboldened!"

Resource: Go to www.dankennedy.com to find out more about how deal flow is king.

Do You Ever Wonder What Your Team Really Thinks About You?

Like it or not, it starts with the owner. Here's a list of what one of my private clients has been wrestling with after asking his team what they thought of him. How do you measure up to what your team really thinks of you? Always a really good question to ask, especially for us move-at-a-million-miles-an-hour people!

- Open-minded to change, but unapproachable
- Good with outside activities
- Sometimes not receptive to new ideas
- Not bending on closure ratio

- Takes money out of my pocket
- Versatile knowledge
- Refrains from recognition
- Be in front office more
- Uses 20 words when three will do
- Perceived as inaccessible
- Father figure
- Seems burned out
- Extremely smart
- Says one thing and does another
- Knows the numbers
- Not around very much

Three Steps to Synchronize Your Marketing and Phone

Step 1: Get your marketing right to make your phones ring to generate the leads you want and the traffic you need.

Step 2: Get the phones right. Turn them into automatic phone sales opportunity machines that secure appointments.

Step 3: Get the sales mood, Sales Talk™, sales language right. From the time your prospective and established customers approach your business, revamp your business so that every nook and cranny has been transformed into the sales mindset. Whether by phone, in-person, at the reception desk, your outsourced call center — all must be covered, all must get it and all must be one. Everyone that represents you and your business is in sales.

By the way, this doesn't mean that you're a schmuck or you don't care anymore or that you don't want to help your customers or that you're dishonest. Not at all.

It means that no matter how you feel inside, no matter how important your career is to you, you also know that at the end of the day all that really does matter is what did you sell? Because without a heavy hand and focus on sales, you have no business, no check and no wealth.

Here are some horror stories about what really happens when businesses don't pay attention. Learn from my real-life experiences...

Accountability

This is about how your staff can run the asylum, make their own decisions about how they're going to do things if you don't monitor them, and in some cases, micromanage and stay on top of them.

I recently visited a large electronic retail chain, you know, where they usually answer the phone with, "Thank you for calling XYZ. You have questions, we have answers." It might be a lengthy greeting, but they always say it like clockwork. Lately they've been answering the phones with, "Thank you for calling XYZ. How can I help you?" I said to the retail salesperson, "Don't you usually say 'you have questions, we have answers?'" He said with a smile, "I think I like it better when we're saying what we say right now." You see, they'll make their own decisions on what they want to do and when they want to do it if you are not on top of them micromanaging, watching every detail and coaching ongoing weekly sales drills.

There is absolutely no way you can make your telephone work in your business if you, as the owner, are not on top of your staff. Very specifically, you need to be able to hold them accountable for goals you want them to have on a daily, weekly and monthly basis, so you need to set specific goals that can be measured.

For example, with the inbound calls, starting with you as owner and trickling down to any other levels of supervision that you have, you have to make sure you track the inbound calls, know the dates and times of the calls, what kind of call came in, what was the action taken on that call and who has the lead now. Where did the lead go once the call came through? That includes voice mail and answering machines.

So hold your team accountable for tracking calls on a regular basis and know what happened to each of them. Have weekly sales drills meetings communicating with your staff having them recite (in the sales drill, same agenda every week) what they did last week for scheduling appointments. "Here's my goal for the upcoming week, I plan on closing x amount of appointments

Resource: To learn how to automatically track your calls, go to www.Great BottomLine.com.

and here's how I'm going to get there." Everybody recites that; there is no excuse and you must make it happen in your business.

If you think you can just outsource your training, mystery calling and coaching, you're mistaken and wasting your money. You might as well not do the training program.

With regards to your marketing, what's working or not working to get your phone to ring? You can't really measure your marketing if you don't know what's going on with the phones. If you're in a situation where you're a practicing owner and you don't currently have a system in place for communicating this way with your staff, you can begin now. You don't have to make it perfect to execute it, you just need to begin and teach your staff that you're moving forward and changing things.

> **Resource:** Go to www.ChrisMullinsConferencing.com for free conference call services. You can even have it recorded and replay it for ongoing staff training.

Staff — Internal Customer Service

I was leaving my doctor's office and meeting with one of the nurse practitioners whom I respect very much. She's very good, has a terrific personality, mannerism, etc. It's always a good visit. When I went to leave and reschedule my next appointment (hint, everyone should always schedule their next appointment when they leave your office), I got to see another side of her.

She was communicating with one of the other team members (the person who answers the phone) and they were doing a little bit of early morning socializing, which is okay except that the nurse practitioner is in a position of authority. So when I saw this different side of her, I initially thought she's showing a good personality and socializing with a team member.

What I ended up seeing more was her sort of bringing herself down to the same level as the team member with the type of socializing she was doing. Her body language was very different, she was goofing around but I think it crossed the line a little bit. What wasn't okay was that it was very clear the conversation she was having with the front office assistant was not appropriate. It was more about "today's the big day," "have your battle gear on," "we're going to have performance reviews," meaning not that the nurse practitioner's doing the reviews on that person but each is having their own performance review.

For the customer (me), it makes it seem like there's some internal stuff going on here with whomever's doing the interviews. There was whispering, too. You don't want to have whispering because even if you're whispering about positive things, the customers or customers who are waiting there to do business with you are thinking it's about them. Unfortunately, I'm able to hear the whispering because I'm right there.

Ultimately, I was disappointed in the nurse practitioner. It changed my impression of her because one of the only reasons I continued to go to this doctor was because of her.

One of the things I often talk about is you can't be so relaxed with your established or prospective customers, whether on the phone or in person, that you let your hair down. When you do that and try to become their friends, all kinds of problems can occur.

Now, when I'm leaving, I'm not thinking as highly of her so my respect level has changed. Maybe I won't schedule an appointment or show up for one. Maybe if there's a different treatment recommendation, I won't trust it. While I was making my appointment across the room with the scheduling person for December, I observed that it's almost the end of the year. Her response was, "Yeah, I'm stuck here. I don't get a summer." I had always thought highly of her as well, but she's let her guard down a little bit too much and you just can't do that. It kills all aspects of your business and you have to monitor it on a regular basis. Continually talk to your staff about examples such as this one. Bring it to the surface, talk about it and have solutions for how to avoid it.

Here's an example of why it's important to monitor your team, listen to phone calls, record phone calls and do ongoing training and critiquing in all aspects of your business. By the way, having a script will be perfect for this.

Not too long ago, I called a large hotel chain (I stay there frequently) to make a reservation. I do that and they always ask for my rewards number. I never have it on the top of my head so I ask them if they can get it. They've already trained their team to answer the phone a certain way, which is good, and ask for your rewards number, which is great.

So I'm making my reservation on the toll-free number, their call center. One quick thought here: Even if they didn't have their information in front of them, they should assume "what other reservations may I make for you while I have you on the phone?" (So here you are making a reservation for hotel A and perhaps you didn't say you have another one, and maybe you don't, or maybe you will six months from now.) If they prompt you to think about it, they're helping you because you'll be doing it all at one time; it's convenient and you don't have to call back. And the hotel looks good doing that while helping to bring in more business.

When they ask for my rewards record, they should look at my records and see if I have any points, bonuses or specials coming to me as a result of being a rewards member. Not one time when I've called to make a reservation have they said, "Wow, Ms. Mullins, you are just 5,000 points away from x, so please continue to stay with us."

Here's a good example. As I'm on the phone making the reservation at the call center for one hotel I ask them, "By the way, do I have any bonuses or benefits coming to me as a result of my rewards number?" They took a couple of minutes to look and it turns out that the reservation I made ended up being free. There was only one thing I needed to do because I was 1,000 points away. I had to buy 1,000 points for $12.50. But you know what? If I hadn't asked the question, it wouldn't have been brought up. Now maybe it wouldn't have been brought up because I wasn't at the total points to have free rooms and I had to buy them, but they could have informed me without me having to ask.

(The reason why this is good is because they're saving me money, but this is making more money for the hotel because it's using their rewards program and they're guiding me on how to use it. Now I'm going to work really hard to make sure that all the places I stay are at this hotel. This is a good example of paying attention to the details, of upselling, of understanding how to use your computer system in all aspects of your business and getting more scheduling.)

After that I said, "Listen, while I'm on the phone, I need to make a couple other reservations and they're at your hotels. Can you do them?" So they did.

Here's another one. I had another reservation to make and it wasn't the same hotel, at least it wasn't obvious to me. It had a different name. So I had to find this place and get the number. I called and the person answered the phone great. I was very impressed with one of the things she said: "Let me be the first to welcome you." For those of you in the Chris Mullins' Phone Sales Training and Critique program, I teach you that. In fact, many students have difficulty saying that, they think it's too mushy or cutesy or something.

So I made my reservation with this hotel and she asked me my rewards number and come to find out they're part of the same hotel chain. The previous call I made didn't even know a particular facility was part of their group. That's dangerous. This is a great example on several fronts: on upselling, on knowing your products and services, on everybody answering the phone the same way, on not just having a good tone and phone etiquette and on being sales oriented.

Team Sabotage?

Here's an example of how your team will make the rules for you. While getting my coffee in the morning, it usually comes with four shots of espresso and I always get the large red-eye with three shots so they charge me for one less shot. I mentioned that to the person who was ringing me up and when she gave me the total, it sounded like it might be a little more than usual, so could she charge me for the three shots.

She said, "Oh, I'm just going to charge you for a small." So instead of listening to the initial request and taking care of it the right way (okay, mistakes happen), when brought to her attention, instead of properly putting through the sale via a credit, etc. it ends up making it easier for her because of her frustration. It's changing the order completely. So you lose money when you don't train and coach your team effectively on a regular basis, or provide scripts or rules or have steps in place.

They make up the rules. That's one transaction, how many transactions a day is this individual doing that loses money not only to the bottom line for the day, but on a regular basis? Because it doesn't sit well with customers when things aren't done the way they're supposed to be; a) for their pocketbook; b) for customer service; and c) it just changes the way they feel about your business.

In addition, my partner had asked for a muffin and they usually have day-old muffins a little cheaper and the person at the register wasn't sure if they were day old or not, so she said, "That's fine, I'll just charge you for a day old muffin." Well, the day old might be $1 and the regular muffin might be over $2 so, again, they make the rules and decisions because how they're hired and how they're trained. It makes the customers feel uncomfortable about your business and they start to wonder what other parts of your business that affect them are being done incorrectly.

One day, I went to Circuit City and what do you think happened? What happened was as soon as I walked in (I knew what I wanted) I noticed a ton of red-shirted salespeople — people everywhere. They have a lot of staff, mostly male, and I ask one of them, "Do you know where..." And before I could finish, he said he'd punched out. Then he turned around and walked away. Someone overheard him say it and told me to go over and talk to this other guy. Well this other guy was busy, so here I am standing there and there are a lot of red shirts standing around. Finally I find someone wearing a light green shirt. I'm with my partner and she walks pretty slow (she has a problem with her legs), but he's not paying any attention to the fact she has to walk slower. I catch up with him, he's not smiling, he's kind of curt and he's not really looking at me; he's looking around and I asked him if he had the product I needed. He said he didn't so I asked about another product and he wouldn't really look. I ended up finding my part after he left but I left not buying anything because the experience wasn't good.

After my stint at Circuit City I went to Best Buy to look for the same accessory. A good greeting for retail would be to say, "What brings you in today?" vs. "How can I help you?" This Best Buy fellow just said, "Hi folks" and walked away. I saw him a little later near the accessories I was looking at and I asked him where my part would be. He walked in front of me, stood in front of the display, held his hands out and walked away. Didn't say anything to me or ask any questions. And that was it. No one else came over.

I did find what I was looking for, but you need to think about your business. Every sales moment you have, you're losing money with all these different experiences. And I'm not bringing them to the table because I'm writing a book, these things happen all the time — you know that. What about your business?

This is about how you inform prospective customers. A little story here. I was sitting at a coffee shop at a hospital waiting for a friend who's having a procedure done and I see the owners and the staff walking in getting their coffee. It's about 6 am in the morning and it's clear to me (since I've been coming to this shop three times a week for several weeks now) that they're bouncing and smiling and seem like they're in a great mood. It's apparent it's coming from the staff behind the coffee counter. They see these team members every day, they talk to them and it's like they're a bright spot in the day. When they're done getting their coffee, they're off to their jobs and nobody knows what they're about to enter. I don't mean just because they're in the medical profession, but we all have different things we enter into when we get to work.

So when they go to get their coffee, they see the smiling faces behind the counter who are consistent, their attitudes are positive, and so for the customer who's greeting them and who's familiar with them, it's a bright spot for them. You, too, influence your prospective and current customers who call you. That can be their bright spot in their day.

Understaff will run the show. They'll make their own rules. An example: I called an inn in Meredith, NH, to make a reservation. It's 11 am at the height of vacation time. After about six rings on a toll-free number, someone answers the phone, "Good afternoon. How can I help you?" I said I'd like to talk about making a reservation for a few nights and he says to me, "OK, well, I just walked in the door. If you can hold on one minute, I'll boot up my computer."

First off, I don't care if he just walked in the door. No customer cares about that. Whatever is going on behind the scenes, fake it. The only reason I let the phone ring is because I had been there before and knew they were open. But any new person unfamiliar with their business will not let the phone ring and be as patient as I am (it's already been over a minute and I'm still on hold). The other thing is, I'm writing a book and I'm a teacher of correct ways to handle the phone, customer service and improving the bottom line with your telephone. So I use this information as material. Simply put: Other people are not going to wait.

An important point I want to make is the entire sales process you have at your business is critical. It's critical for you to identify — not have a knee-jerk reaction of going forward with your day-to-day business — your sales process and to outline it and keep an operations manual on it, if you will, for people to refer back to.

The sales process has to be against your marketing but it also has to do with from the time the phone rings. As soon as your phone rings, it starts the sales process. The person answering the phone is a crucial individual who can collect so much information, not just by language and words, but by the feelings and reactions of the callers. All of that information should all be documented, noted on your scripts and passed to the next person in line. But the biggest breakthrough for your business with regards to the sales process and the phone is to have a weekly sales drill where you're all talking about each case and each customer.

All of that information, every detail, needs to be passed on to the next person down the line; everyone who has a point of contact, communication moment or a sales moment with your customers. But you also want your staff to have their own have sales moments even if they're not directly communicating with that customer. So, if Mary is in the waiting area to see the owner, but she's not going to be meeting two or three other people in your business that day, everyone in your business should know Mary and why she's in and go out of their way to say hello to her.

Running the Asylum

In a lot of situations, staff just doesn't understand your business or your prices and services. Unless you look over their shoulder with ongoing training, you don't know what's going to happen. Here I am at a very large chain store (and this is across the board with every retail store). I was looking for a power cord for my MP3 player and the fellow working there said they didn't have one for my type. So I look a little further and there, right in front of his eyes, I find it. But he didn't really look or try. It's the same thing that's happened in other retail stores and it's business that they're losing.

I was also looking for some envelopes and I asked the store manager where they were. He acted like it was a bother to him and he brought me over and showed me something but he didn't work very hard at it.

The message here is you need to make sure your staff continually has a script on how you want them to perform with inbound calls, outbound calls and all aspects of your business. Then you need to do mystery calls, pop quizzes, etc. Check to see if they understand. Not because you want to micromanage in an unhealthy way, but because you want to improve your bottom line.

How to Schedule Appointments

I got my hair cut at a salon and here's what I observed while I was waiting. After a customer was done one of the employees asked if she wanted to schedule the next appointment or not. She actually said "or not." The customer was hesitant and said she didn't know and could she just pop back in sometime. The employee said sure, no problem, you can just pop in.

The correct way to respond, since you want to continue the relationship, would be to say at the end, "You're next scheduled appointment would be…" Make it for the same day the customer is in there already and the same time frame. Assume it's going to happen that way — don't ask them if they want to.

If the customer says they might be busy or will call to schedule or pop in, say, "That's okay, we'll keep you on the schedule. Here's my card and if you need to call and reschedule you can do that."

There were also two other employees in the salon who were very different than this one. They were very professional and were more sales-oriented. They automatically said when the appointment should be and did exactly what I just explained; mentioned the day and the time and dealt with the changes that might need to take place after the customer said something.

On another visit to the same salon, I noticed there are things that happen and the effect it has on the receiving side. For example, while I'm checking out, the person who did my hair showed me some products and asked if I'd like to buy any of them. Kudos to her.

However, there were a couple of things that could have been done differently to really cement in my mind the importance of the products. Customize your presentation to me. Why did you use these products on me? While using them, what did you notice and learn from me and the questionnaire you had me fill out that was important enough for me to consider buying these products? What will happen for me?

Remember, it is a sales presentation, so providing it at the checkout counter while I'm paying and there are other people in line isn't such a good idea because it diminishes the value of the presentation and the importance of the product. The person who was doing this with me was a little distracted. One of the things that affected me as a customer as well as a professional is she got distracted by another person that had just entered the salon and lost eye contact with me a few times. That told me she wasn't as interested in me at that moment and was kind of disqualifying me through her behavior. That's a good reason to change where you do your presentation.

Of course, that's in person, but even on the phone you have to make the front desk person so important that they can focus on what they're doing, which is presenting an appointment.

There are so many sales moments and opportunities that can add value to what you do, but can also hurt you if you're not very careful.

AHA Notes:

Chapter *Five*

Embracing the Telephone

Shocking Facts!

Most business relationships rely on the telephone, and how you sound makes up 84% of the message you send. When a prospect or customer decides to finally get around to calling you because of the promotion you sent them in the mail, they want to talk to you, the live person!

Your answering machines or voice mail can't sell them, or in the first few minutes say, "Wow, you called the right place!"

More Shocking Facts

Eighty percent of NEW customers will NOT call back or leave a message when they get a voice mail or answering machine.

Only 25% of business owners even believe the telephone has a major impact on their business (even though it is the only point of entry for all new customers).

Some Statistics

Many people think that communicating effectively is merely a matter of finding and using the right magic words. They believe that using certain words in the right order will get the results they want.

Scientific research tells us that attempting to persuade by words alone is about as effective as trying to chop down a tree with a Swiss Army Knife.

To be a truly effective communicator, your body language and tone of voice must be consistent with your content. Even the most powerful words spoken in a monotone with lifeless body language will fail to rouse anyone. In a study conducted at UCLA, Dr. Albert Mehrabian found that when verbal, vocal and visual signals are inconsistent, content counts for a mere 7% of the overall message. In such a situation, 55% of the message is transmitted by facial expression and body language; and 38% comes from voice quality — pitch, tone, volume and inflection.

So, when you think about it: If you've been given the luxury of "face time" with a person you want to influence, a warm, friendly smile, a firm handshake and good eye contact can work wonders. But if anything about your voice is flat or distracting, annoying or boring, you've just reduced your effectiveness by 38%.

And how often do we spend face-to-face time with our customers these days of phone-computer-fax? Discounting the attractive physical impression, what counts is not only what you say, but how you say it.

Keep in mind when reading statistics like this that answering machines and voice mail to capture new leads from prospective new customers can't possibly convey the message that you're here for them, that you want their business and that they absolutely called the right place, so come in for the appointment.

You've already purchased a phone system. That money is spent. You've already hired staff. Most business owners currently advertise in the yellow pages, newspapers or special direct marketing programs of some kind. The primary goal for all of this spending is to get new customers and they all come through the telephone!

Do you know how many new customers you lost this month?

You're losing thousands of dollars each month one call at a time. Most business owners feel and say it's not happening to them, their business is different; this is how they have to do things even though some have the Chris Mullins' Phone Sales Program™ team looking over their shoulder every month proving that their phones are being answered by machines and not by live people.

POWER NUGGET™

What do you have to do to fix the problem immediately?

The smarter question to ask yourself is what do we have to do to fix this problem immediately?

You'll get more ideas when you think like that verses "our business is different." Unfortunately, we sometimes need to use voice mail. Another question is how much are you spending each month on marketing to get your phone to ring? That alone should be enough reason to say we need to take what we're uncovering with our phones in our business to fill the holes in our bucket, fast!

We need live people on the phone to give every new prospective customer what they want — a live person — so let's provide a solution and test a few ideas to see what works best.

How do you know that the messages you are getting are being captured?

Don't put this off! There are few concerns more pressing than this one, because every new lead, every referral, every single dollar that you spend on advertising is driving new customers to one place — your telephone!

All the phone training in the world won't help your staff to set new appointments if they don't answer the telephone when the call comes in.

We would all like to think that new customers would conveniently leave their name and number when they get your voice mail, but the fact is it rarely happens. Existing customers and people trying to give you money might leave a message (they won't like it), but a brand new customer will most likely call somewhere else.

When they finally tip over to make that call, they're ready to talk to you — the LIVE person! You've got to be ready to say, "You called the right place. I have an opening today at 10:00."

When someone calls you after business hours, they are not really expecting you to be there. When they call at night, they were planning to leave a voice mail. On the other hand, when they call during business hours, they are expecting to reach you.

So, for many customers, they're conditioned, and therefore expect, to reach you during 8:30 am until 6:00 pm. These hours must be covered by a live person. Do you know how many hours per week your phones aren't covered by a live person?

What about the prospective customers that call your business during lunch break? Are you answering the phones during those times?

POWER NUGGET™

Businesses that use their telephones... must focus on their telephone expertise

Businesses that use their telephones for the main point of entry to their business must focus on their telephone expertise just as much as they do with the technical aspects of what they do and the time and money they put into lead generation marketing.

Even More Shocking Facts

Here are thirteen top excuses why your phone isn't being answered by a live person. Here's what we (the Chris Mullins team) hear from students and members (your staff) across the country on why they couldn't get to the phone and/or why it went to voice mail.

1) I was with a customer, no one else picked it up, I'm only one person.
2) I'm trying to a hire a new assistant to answer the phone and schedule appointments. I can't hire someone until I get the business owner to tell me to go ahead and put the ad in the paper. That was four days ago, now, I have an open house to attract new customers but no one to help with the phone to make appointments and give directions for the open house.
3) I was in the back taking a break, didn't hear the phone.
4) I had a doctor's appointment; the person that's scheduled to help me didn't show up yet, but I had to go.
5) Everyone is busy with other customers.
6) We don't have call forwarding.
7) I'm new and just thought someone else would pick it up.

8) I had the day off.

9) I couldn't get into work because my car wouldn't start.

10) I got a call from the daycare and needed to pick up my daughter.

11) I had to have lunch, it's my time, if I don't take it I don't eat.

12) I didn't know how to take the phone off the answering machine from the night before.

13) We don't have voice mail. If I'm on the phone and someone else calls, my phone beeps, but sometimes it's hard to take that call because I'm already concentrating on the person I'm speaking to, so hopefully they'll call back.

Ray Kroc famously correctly identified clean bathrooms at McDonald's as marketing. Walt Disney said the same about clean streets at Disney.

This holds true for the telephone! The telephone is the key to your business the first point of entrance (voice) in your business. The telephone will make the business owner look good or bad.

Arghhh! Stop Putting Callers on Hold

Especially before you even meet them, greet them, get their name or know why they're calling.

I personally call clients to check in on their phone skills. The other day I was on hold for three minutes with a client. Please, this is not acceptable!

If you're a dentist, your average new patient (customer/sale) represents $1,500. Doctors seeing four month cases = $20K+ per new customer.

I can tell you this: If you're putting your callers on hold like I just explained, then you're losing business — period!

Don't you feel that way as a consumer? Doesn't it tick you off when it happens to you? Fix it! That doesn't mean, "Chris, we've always done it that way or we have no one else," etc. It means you're losing hundreds or thousands of dollars.

Let's step way out of our comfort zone and just decide to fix it. Then, and only then, will you come up with tons of solutions you never thought of before.

Eight Instant Stress Busters in 20 Minutes at the Front Desk

Even in the middle of a stressful day, you'll fair better using a few strategic destressing exercises. These real-time coping skills comprise a holistic progression from body to mind to spirit. Choose the best ones for the moment, or complete the whole routine within 20 minutes.

Stretch: Feel better immediately – without even getting out of your chair – by slowly bending forward, leaning back and twisting your spine in both directions. Standing, slowly reach for the ceiling inhaling, then bend and reach for your toes, exhaling.

Progressive Muscle Flex/Release: Tense strongly from feet up, and hands inward, while inhaling through the nostrils. Release suddenly, from head to toes, exhaling through the mouth.

Self-massage: For stimulation and relaxation, rigorously massage your head, neck, face, hands and feet.

Eye Palming: Rub your hands together vigorously until hot; place them over your closed eyes for soothing energy.

Breathe: Squeeze all air out from the belly up, then inhale, fully expanding abdomen, rib cage and upper body. Pump the breath in and out forcefully from the belly to stimulate internal glands and organs. Feel the breath passively without controlling it.

Relax: Elicit your "relaxation response" by silently or audibly repeating a word, sound or prayer, ignoring all other thoughts. This internal focus alone can reduce muscle tension, blood pressure and anxiety. Focused mind exercises produce body results!

Balance: Lean forward and back, then side-to-side, reducing movement until still and balanced. Center your awareness deep within the belly.

Get Grounded: Feel your connection to the earth.

Autogenetic programming: Repeat these positive statements in first person, present tense:

- My forehead feels cool
- My limbs feel heavy and warm
- My heartbeat feels slow and quiet
- I feel completely relaxed

Source: Lonny Brown

Get Out Your 3 x 5 Lime Green Index Cards!

Here's a list of psychological triggers many top-performing entrepreneurs have in their toolbox for success. Shouldn't they be in yours?

Keep your 3 x 5 lime green index cards in your pocket, on your desk, car dash, kitchen cabinets, bathroom — you name it. The idea is to put several of them in the places you frequent so you will see the words over and over again.

FMF – Feared Most First

Do what you fear most first. I learned this from Mark V. Hansen, publisher of *Chicken Soup for the Soul.*

What's on Chris Mullins' Kitchen Cabinets?

They're full of lime green 3 x 5 psychological trigger index cards...

Time Blocks

Add Minutes to Exercise

Make Coffee Night Before

What You Focus on Expands

Victory Book

20 Goals 20 Minutes

Answering Service Nightmare!

Do you think you're safe because you're using an answering service? Well, you're not! They still need to handle the phone correctly, but they don't — not at all! They represent you and they represent your company. Even though you're outsourcing you need to think of them as your staff, therefore you need to take hiring them seriously.

I recently sent this note below via fax, voice mail and e-mail to the president of a real estate investing company. How did I find out about this problem? The president wanted to discuss a project with me so he asked me to call him. I did and here's what happened.

[Don't think for a minute that because you're not in real estate investing that this doesn't happen to you. It does and it happens even with the popular services that we're all using right now. You absolutely must inspect what you expect — no exceptions.]

Urgent!

So, I hear laughing on the phone as soon as the answering service picks up; no greeting, no company name, I couldn't understand what they said and I really thought I had reached someone's house!

I ask for the president by name and they immediately (didn't acknowledge me) transferred me to an automated line that tells me (with no greeting) to punch in the voice mail box number.

Obviously, that's not for me and the business card I was given to call this person doesn't have it listed either. I hang up and call back (the only reason I did so was because the president of the company had asked me to contact him. Otherwise, I would not have bothered, which means *lost sales opportunity*).

I called back and this time the phone rings off the hook. I wait because I just called (your prospective clients wouldn't wait and they would hang up). Finally, the service comes on, I ask if I'm calling the right place and she says, "You're calling an answering service."

Okay, but am I calling the right place? I say the name of the company, repeat the number and the name of the president. The service replies with, "Yes, but he's not here, you can call back." I ask, "Do you know his voice mail extension so I can leave a message?" "No I don't, call back." "Can I leave a message with you?" "No, you need to call back."

This is very dangerous and, I kid you not, very common! You may have decided to outsource your calls to a service so you don't have to have staff and you may have one of the "popular services" that others have been using so you may think you're safe, but you're not! Whether you're using the popular guys or not doesn't mean they know how to handle the phone. They need monitoring and training about you and your business, not only on the right way to use the phone, but how you want them to use the phone to represent your business, your leads, your bottom line!

Of course, the business card had no other number or e-mail to contact the individual — another problem for clients, prospects and the folks that you want to contact you.

Business Owners: Pay Attention to the Details!

Get it right. Just because you're spending time and money on marketing to get the qualified leads to get the phone to ring doesn't mean you can forget about the phone details.

Use the same energy you have for marketing with your telephone. They go hand-in-hand and should always be part of the same discussion when it comes to the bottom line, marketing, staff and outsourcing.

My own business now has a virtual inbound and outbound call center that represents Mullins Media Group™, LLC. During the development of this call center I made sure that I personally interviewed and trained the entire team and owner before I decided to create the partnership.

Part of the deal was that I won't put the time into interviewing team members unless we agree up front that the owner and team will be trained in ongoing teleclasses by me on how we do things for our clients with regards to the telephone.

Resource: Go to www.GreatBottomLine.com/marketingvideocoach for solutions to answering service nightmares to get a complimentary consultation with Ron Sheetz on how to get your phone to ring. Go to www.UniversityOfPhoneProfessionals.com to get the cheat sheet on answering services.

Putting Callers on Hold

Chris,

I have a question in regards to putting the caller on hold. We are VERY short staffed at this time. I am the only person answering the phone (five lines) as well as booking, processing customers, answering questions in regards to customer treatment, harassing insurance companies and collection calls. We are in the process of hiring but for now it is just me responsible for seven providers and four hygienists.

When I am processing a customer or discussing a financial arrangement (making the sale), I feel THAT person deserves my undivided attention. You were saying that we should try to get details when we pick up the phone before putting them on hold, but for me, the moment I do, THAT caller gets all my attention (which they deserve). If I have a customer in front of me, I don't feel it's fair to them to begin something with a caller on the phone.

What is the best way to handle this? Until we get additional staffing, would it be better to let our answering machine pick up if I don't think I can give 100 percent attention to that caller or do I ask them to please hold?

Aileen

Aileen,

Get temporary help from an agency to answer the phones. Get anyone you can to help with answering the phone even if they just take the name and number for you to call back — a friend, previous high school intern — and be sure to go over your script with them.

In the future, plan for this, know how you'll fix it and put the system in place now. Yes, the person you have on the phone at the moment is most important, especially the new customer (sale), and is critical.

The doctors in your business must know about this, you need to make it your mission to put this problem right in front of them as often as necessary to get the point across that YOU simply can't do business like this anymore. Do the math for them, that should help.

Sample call: What are your hours (or, whatever the call is for)? May I ask your name? Mary. Mary, would you mind holding for a moment? I'll be right back to answer that question. Wait for the REPLY. Then, go.

Really, this doesn't take more than a moment. Many times our panic and frustration is what's overwhelming us more than anything, thinking that this is taking too long.

Or, if you're in the middle of another conversation and you have no support. After you get the caller's name say, "Mary, that's a good question, what number are you calling from?" 123-4567. "I'd like to call you right back as soon as I finish with the customer I'm talking with now. How long will you be at that number?" Thirty minutes. "Okay, I'll call you right back."

Note: I do not recommend putting callers on hold that are new or prospective customers, as you know, but your note forces you to do so, you've got to quickly take care of the problem at hand, stop the bleeding. However, I strongly recommend you get some other live bodies to answer the phone for you and take messages if necessary to save you from putting them on hold and having to stop working with the customer you're dealing with at the time.

Finally, if you can't find anyone as mentioned above (a temporary employment service is best) to answer the phone, take messages, etc. then you have no other alternative; you have to let the calls go to the machine.

Make sure the machine says something very personal and friendly, not the typical flat message. Sound like you're very disappointed that you couldn't get to them personally.

Chris

The Mirror Technique

Many of you have heard me talk about using a mirror at your desk to remind yourself to SMILE before you pick up the phone, to stay focused on taking care of each call like it was the first one of the day, to shake off any bad experiences you might have just had before the telephone rang.

> **Resource:** Read about the mirror technique in *The Magic of Believing* by Claude Bristol and how he uses it in his own businesses.

Read On

In my old organization, where we had to make a complete about-face to avert disaster, I first used this technique by placing a mirror in a back room of the offices where the employees left their hats and overcoats. It was placed so that everyone had to see it when entering or leaving the room. At first I pasted strips of paper with slogans such as "We're going to win," "Nothing is impossible to an indefatigable mind," "We've got the guts, let's prove it," "Let's show them we're not licked and then go to town," "How many are you going to sell today?" Later we took soap to write the slogans directly on the face of the mirror.

Every morning a new slogan appeared, with the sole purpose of convincing our employees that they could get business even though others were struggling to keep their doors open. Eventually mirrors were also placed alongside the door frame of the main door to the office so it was the last thing salesmen saw as they left and more alongside calendar frames on the desks of all salesmen and executives. The startling thing about it was that during the worst of the "depression" days, all the salesmen quadrupled their income.

Chris' BIG Question

How far will you go? Will you put mirrors up in your own business and explain to your team what the mirror technique is all about?

Expert Advice – Dialing for Dollars – Outbound Calls

Does this sound like you when you're calling your customers for any reason at all?

Sometimes salespeople will make calls because they're supposed to, not because they want to. They have a negative attitude and a feeling of "I'm just trying to get this done and out of my hair," and the customer or prospect can sense it. They're just going through the motions with no enthusiasm.

If they're professional salespeople (which is the mindset you want), they'll have financial goals and should know the exact value of their time and the value of a new customer. They need to ask themselves, "If this is the value of a new customer, why do I have a bad attitude when I call them on the phone?"

Ask yourself why you're calling customers in the first place. Do your homework before you call. Know why you're calling them, have a script to keep you on track, practice your presentation and believe in what you're doing. If you can't sell yourself on why you're making a phone call, you can't sell them.

Know what your strengths and weaknesses are when you're talking on the telephone. If you tend to get nervous, practice and learn to relax. Practice a script and record it into a tape recorder. Then listen for where you can pause, take a breath or just relax. Call a friend and role-play over the phone.

Build your self-confidence. Know your business, services and products backwards and forwards. Learn all you can about the marketplace and your customers. The more you know, the more reasons you have to call and the more you can offer, simply because you know more.

When you talk to customers on the phone, concentrate on painting a visual picture for them. You want them to be able to see what you're selling through your words. If you really know what you're talking about and are comfortable selling over the phone, your words take on visual meaning.

Build and/or enhance a relationship over the phone. Call with information important to your customer, share something personal or even just wish them a happy birthday. Customers will pay a premium for your product or service if they know, like and trust you, and a phone call can reinforce that relationship.

Listen for what the customers' real wants are — really listen. Ask them to repeat something if you don't completely understand what they're saying. It pays dividends.

Know exactly what you want to accomplish from your phone call and get a "call to action."

The Missing Link to Attracting More Customers Uncovered!

Q: Do you know what the missing link is to attracting new customers and referrals to your business?

A: The telephone.

Q: Do you know what the most valuable asset is in your business?

A: Your customers.

In a study conducted at UCLA, Dr. Albert Mehrabian found that talking on the telephone, the actual words you use, account for only 16% of the way you and the products and services you represent are perceived.

The remaining 84% of your impression depends on the sound of your voice and the feelings people get when listening to you.

This is important... How do you know when people are ready to buy right now? You don't! Which is why you've got to stay in front of them with effective marketing.

What happens if you have awesome marketing but your telephone skills are poor? Answer: No new customers.

In addition to effective marketing you've got to make sure that when the phone rings you have top notch professionals answering the phone, ready to close new customer appointments and referrals.

Let's say you want just 20 new customers each month. All your marketing is done so that the phone rings with new customer opportunities and referrals. What? You think that referrals aren't in the same category as new customers? Not so.

Just because they've been referred to you doesn't mean that they'll give you their money. It doesn't even guarantee that they'll call, but when they do, you've got to be ready.

Back to the Phone

Your marketing is now working and the phone rings with lots of new customer opportunities. That's great, but it's not enough. It's time to focus on your staff; everyone that answers the telephone. Get them excited about the phone each time they hear it ring. Get them motivated and looking forward to closing new customer appointments.

Coach your staff on perfecting basic telephone and sales skills. Be sure they understand what sales is, how to answer objections, how to use a script without sounding like they are, how to close referral appointments, how to listen and how to follow up.

You can't go wrong with coaching your entire team on telephone skills, sales and customer service relationship marketing.

"Sales are made everyday. Some will close — you or them. What's it going to be?"

AHA Notes:

Chapter *Six*

Staff Recognition

Improve Your Bottom Line With Peer Recognition Programs

Everyone has a basic desire to be appreciated. Just a simple "Thank you, well done" can make their day!

Did you know that the lack of recognition and praise is a major reason why people become dissatisfied with their jobs [often ranked higher than pay or benefits]? I'm sure if you think about it just for a moment, you'll understand how this would happen.

Peer recognition programs are easy to implement and there's nothing quite like getting recognition from a peer. Programs like this done right actually carry much more value to the individual getting the recognition than when they get kudos from the boss.

The idea is really twofold: 1) catching people doing things right and 2) getting employee praise. However, praise must be specific, not general. For example, "Wow, you're really great!" doesn't qualify as much as, "Wow, Mary, I was really impressed when I heard you say to your customer that you wouldn't hesitate to buy the same [fill in the blank] if you were in the same situation."

Being SPECIFIC says that you really did listen, you heard something that was very important and the receiver realizes that what they said, for example, was very important and therefore will want to repeat it again.

Recognition programs offer a way to create a positive workplace where employees take time to observe their coworkers' work habits, notice their efforts and follow their positive examples. Some examples of recognition may include...

- An act of kindness or personal support
- Consistent dependability and promptness
- Exceptional performance for complex task
- Working extra hours or on personal time
- Insightful contributions to drills
- Motivating, encouraging team spirit
- Delivering excellent customer service
- Offering solutions and resolving problems
- Service to community and family
- Anytime someone makes an impression

Recognition programs positively impact recruiting, retention and business culture. A simple act of employee recognition can create an environment where they feel appreciated for their contributions.

When employees feel appreciated, retention improves and retention demonstrates company loyalty and promotes a positive business culture which attracts good employees and customers.

What goes around comes around, and peer recognition can keep that positive cycle in continuous motion.

Annual programs are good when it comes to recognizing significant milestones in a career with each individual area. But 365 days is a long period to wait to be recognized for your efforts. Peer recognition presents an opportunity to pass along day-to-day employee recognition and promotes continuous job satisfaction.

Be careful though, don't just throw together a program. You need to think it through. This program should be an essential part of your bottom line and monitored carefully.

Peer recognition programs have a positive effect on productivity, punctuality, attendance and commitment to the business.

Recognize – Reward – Motivate™ (RRM)

Peer recognition gets the best ROE (Return On Effort)™ – Create a recognition program that's company-wide. When an employee catches another employee doing something right, they fill out a "kudos form" which goes in a ballot box that has been left in several different areas of the office.

At the end of the month a team of designated managers goes through the forms and picks about a dozen for each area. The kudos forms along with the certificates get posted on a board. The certificates for catching people doing things right goes to those individuals at the end of the month.

Recognition During Sales Drills

As the business owner and office manager you need to begin a daily schedule on your organizer. Catch someone doing something right and get yourself in the habit of doing this every day with everyone; it doesn't have to be huge.

For example: You notice a team member helping another, smiling more, closing more new customer appointments, talking differently on the phone, handling criticism differently, trying harder or checking in with customers waiting to be helped in your waiting area. You need to do this with everyone. Soon, it will catch on and your team will do it with each other. For now, focus on YOU, then share this concept in your weekly sales drills.

A Quick, Helpful Guide for Your Weekly Sales Drills™

1) Take a look at your notes from your monthly Phone Sales Training Program and phone critiques to put together a list of what's most important to discuss in the sales drill.

2) Be careful with any criticism that you want to share because it can be difficult for many folks. The *delivery style* and the *person* sharing the information is important to consider. This step can make or break the very important lessons to be learned. You know your folks, so you decide.

3) Weekly sales drills are all about sales, it's a bottom-line meeting about the numbers, the math. Discussing possible closing of cases for that week and the month, etc. You also use this time to share successes with everyone bringing an objection they're having difficulty with to the sales drill. They can all jump in and role play for the right solution. Example: Incoming calls as a result of your lead generation efforts, number of scheduled appointments, number of shows and number of new customers.

4) The TONE of this weekly sales drill needs to be positive, upbeat, how to, focused, educational, team action oriented and short (no more than 45 min. (shoot for 30 once you get going with a few). No food at this meeting; it's business, all need to be focused.

5) Topics – discuss at least one topic that has to do with improving your bottom line, including pieces of my inserted comments/lessons in your critiques. Be sure to focus on PAS (Problem, Agitate, Solution) with each customer. You also want to do role playing to practice your script, objections and difficult phone situations.

6) Do NOT include any housekeeping during this time. Schedule it for the same day and time each week. Pick a time that you feel would work mostly with your schedule, then schedule it every week in your organizer and have everyone else do the same. (I recently heard that Tuesdays are good for this. To learn more, go to www.GreatBottomLine. com.) A team can help improve the bottom line just as easily and quickly as they can *hurt* it with poor attitudes like the "hey, it's just a job" mentality.

Keeping Staff Motivated

Chris,

We've seen a significant increase to our sales goals. How can we keep the staff motivated to meet these sales goals when we are behind in previous months?

Sandra

Sandra,

As leaders (owners, office managers), you must believe it can be done. If you don't, you won't get there. One of the biggest responsibilities of leaders is to get the team to trust you, and that everything you say and do is for the benefit of getting them to the end result — meeting sales target. If they trust you, they'll push for you even if they wonder if it can be done. They won't let you down; they'll believe in you.

Chris

Again, you must really believe it and act it with your words, body language and lessons. All owners and office managers (leaders) must remain very positive; have an "it's not over until it's over" mindset, use positive language, turning all negative language from others quickly into positive. Constantly recognize individuals daily; catch people doing things the right way that will motivate them to want to do more of the same. They'll search for other things where they can get more recognition.

Coach and monitor all individuals on a weekly basis. Coach/monitor first the folks who are behind with goals; give them lots of attention with specific, short, how-to steps on what they can do differently to improve. Show them how to do it; search for areas where they may be doing well and sharpen the axe. Search for past successes, lessons or contests that worked. Then repeat it! Push forward every day.

Resource: To find out more about Chris Mullins' Do It for You Call Monitoring and Tracking program, go to www.GreatBottomLine.com.

Worth Repeating: Weak is he who permits his thoughts to control his actions; strong is he who forces his actions to control his thoughts.

Each day, when I awaken, I will follow this plan of battle before the forces of sadness, self-pity and failure… capture me.

- If I feel depressed, I will sing.
- If I feel sad, I will laugh.

- If I feel ill, I will double my labor.
- If I feel fear, I will plunge ahead.
- If I feel inferior, I will wear new garments.
- If I feel uncertain, I will raise my voice.
- If I feel poverty, I will think of wealth to come.
- If I feel incompetent, I will remember past success.
- If I feel insignificant, I will remember my goals.
- Today I will be the master of my emotions.

– Og Mandino, "The Greatest Salesman in the World"

Mind Power

Christopher Bergland describes how to develop an "athletic mindset" in *The Athlete's Way: Sweat and the Biology of Bliss* (St. Martin's Press: June 2007).

Resource: Go to www.Milteer.com for more info on this topic.

Set Goals

The sense of achievement that comes from accomplishing a goal comes from dopamine (a neurochemical). To get regular hits, have management goals that'll give you more opportunities for success.

Get Inspired

Bergland regularly reviews quotes from Muhammad Ali, Charles Lindbergh and Friedrich Nietzsche. "When exercising, one of these quotes will often pop into my mind — usually when I need it most."

Monitor Your Mood

Compare how you feel before and after a fun using a scale of –5 (stressed/cranky) to +5 (calm/elated.) "If you're starting a workout feeling +1, you'll know that when you're done you'll be a +4 or +5."

Use Positive Imagery

Bergland's fridge features Lance Armstrong, the Hawaii Iron Man and Superman — things that motivate him. But his most unique visual cues are rubber bands (sound familiar?). He puts a new one on each morning as a reminder of his commitment. At night, he adds it to a rubberband ball, "a representation of my investment. I take those rubberband balls with me to races to remind me that I've trained hard and am prepared."

SALES LESSON You can do the exact same thing as the above to motivate you each day with your professional goals to improve sales, appointments, attitude, customer service and your personal life. You don't have to be an athlete or a runner to do this — it works for everything in life, whatever you're focusing on as an individual to improve with your telephone skills, number of appointments you close daily or weekly, your attitude on embracing change, integrity, critiques — you name it.

Remember, what you focus on expands. Focus on being the best at what you do — an absolute expert at answering the phone and closing appointments, training or an expert at greeting customers or an expert at sharpening your axe. Remember, it will expand so be sure it's all positive.

A Phone Success Interview

Chris Mullins' Phone Sales Program includes a monthly Sales Hot Shot program. This is just one way we support you by providing recognition and motivation to your team. However, you can easily implement your own internal recognition program, at the very least, catching each other doing something right on a daily basis will set the tone right now.

Read this interview with Emily Neal, customer care coordinator of QA Advanced Dentistry, who was nominated by Dr. McAnally (www.BigCaseMarketing.com).

Mullins: Hi Emily, what's a customer care coordinator?

Neal: An all-purpose front office person who handles the phones, the financials, scheduling customers, greeting and helping when they check out.

Mullins: How long have you been with Dr. McAnally?

Neal: I've been with Dr. McAnally for six years.

Mullins: You've never worked in a dental business before?

Neal: No, I had experience as an office manager working for a construction company, but certainly never with the type of marketing that Dr. McAnally does.

Mullins: Do you feel special now that you're a sales hot shot?

Neal: It's really flattering. I don't know that I'm necessarily worthy of that, but I'm very flattered by it.

Mullins: I remember when I first started working with you — it's only been two years I believe. You were pretty new to whole concept of selling in the business.

Neal: Right. I think the intensity of the selling that we do is a lot greater than anything else we have ever done. Before it was like, if someone makes it into the office, "Hey, let's try to get them to go forward with treatment." Now it's like, "Let's get them in the office" and start with the very first phone call.

Mullins: That had to be an interesting, perhaps even scary, experience for you.

Neal: It was a big learning experience for me.

Mullins: When you think about sales, why do you think you were selected as a hot shot? Setting modesty aside, why do you think you've done so well with the whole sales experience?

Neal: Well, one of my biggest challenges was controlling the phone calls and really letting the customer know that they called us because we're the experts. That was a big challenge for me. It was something I really focused working on.

My goal was to get them in here, to not let them talk me out of it, but to find a way to make them understand that they called the right place and the most important thing is to get them to come in.

Mullins: How difficult was it for you to go from answering the phone and questions to controlling the call and convincing them to come in? How hard was it to cross that line?

Neal: At first, it was really hard. I was pretty easily intimidated, I think, by what I perceived to be confrontation on the phone and someone getting upset with me because I wouldn't give them the quick answers they wanted. It really intimidated me at first, but then I realized I'm doing them a disservice by giving them quick, one-word answers when really what they need to do is come in and see the doctor for themselves.

Mullins: As you know, I have all kinds of students. I have some stars and some who are floundering. Change is difficult for them, but at the end of the day the truth of the matter is that there is nothing I can do. There is only the teacher delivering the right information that works, the tools, the support, the guidance and it all comes down to that individual person and what they do with the information. What is it about you that you were able to go from what you were used to for four years and then change? I mean, you didn't just try the new way with my program here and there, you stayed at it consistently.

Neal: I think there are a couple important factors. One is being positive, because you know a lot of times people will call and they're talking about a very negative situation. By putting a positive spin on it, it encourages them that there is hope. I think another part of it, for me personally, is to honestly have the desire to succeed and the desire to help the patient have success in their treatment. So, if you really don't care whether they come in or have their treatment, it's going to come out in your phone call, in your voice.

Mullins: Let's rewind to what you just said, "desire to succeed." What's that all about?

Neal: I guess it has to do with a lot of different levels. I have a desire to succeed in my short-term curriculum goals. I want to get a certain number of people to come into the office and I want to take a tough phone call and turn it into an appointment when the person is usually resistant to coming in. But I also have long-term goals for the practice. I really get a lot of satisfaction in knowing that I play a big part in its success and that we work together as a team in our office. There are a lot of different levels of success.

Mullins: Right, and you're talking to me about your business part of it and that's very important, but what I want to dig into is you as a person. Maybe your parents, maybe your upbringing; not everyone has that same desire to succeed at what they touch and you apparently have it. That's a big deal. I mean, you work with a lot of people yourself, you can see that not everyone has this desire.

Neal: Right.

Mullins: Many folks can have the thought, "Yes, I'd like to learn something new" and go for it, but to stick with it, that's altogether different.

Neal: I don't look at it that way.

Mullins: What about your parents? Tell me about them.

Neal: My dad is actually in sales; he sells airplanes.

Mullins: How long has he been in sales?

Neal: Thirty years for the same company. My sister is in sales and is a director for large makeup company; she sells makeup. I don't know, I never thought of myself as a salesperson. In fact, I kind of shied away from it because I just didn't think it was me.

Mullins: How long has your sister been in sales?

Neal: Twelve years. My mom is a fantastic stay-at-home mom; she's the backbone for everyone. I have two older sisters. The oldest sister, Mary, is the makeup sales director and my closest in age. My sister Gretchen is actually a stay-at-home mom and my younger brother Adam is a bartender. My dad's name is Paul and my mom is Teresa.

Mullins: You've never done sales before?

Neal: No.

Mullins: What would you say you were brought up to believe?

Neal: To accept challenges and to see challenges as an opportunity to succeed, not an opportunity to fail.

Mullins: Is it really important for you to succeed each day or overall at a particular project? I mean, is it as important to you that when you go home at the end of the day you want to feel like, "Okay, I had a mission today and I was able to achieve it?"

Neal: Yes, I get a lot of satisfaction out of feeling like I've achieved things on a daily basis and even more satisfaction knowing that those smaller achievement will add up to bigger ones.

Mullins: Before the phone coaching program until now, what differences have you seen in Dr. McAnally that have positively influenced you to focus more on sales from the moment the phone rings?

Neal: Dr. McAnally is extremely driven in his own life, that is the underlying theme of our office. We're always striving to be better and so he's very encouraging, but he also has very high expectations that we will put in the effort to succeed. You know he's given us these tools by having you come on board to help us and he's doing everything he can to make the program succeed so he expects us all to take part in that as well.

Mullins: What other types of changes have you seen in Dr. McAnally since we've all been working together?

Neal: Actually, I have seen a change. It seems like his drive is more refined now, so even the phrasing that he uses in consultations with patients, discussing their treatment, is very specific. Before he would just kind of throw it out there on the table for the patient to look at; now he's very specific in guiding them through the process to help them understand the gravity of the situation and the importance of having Dr. McAnally treat them.

Mullins: So you've seen improvement in Dr. McAnally's case presentation?

Neal: Oh, absolutely, yes.

Mullins: How about communication with all of you regarding sales, sales talk, your sales language, sales drills and that sort of thing? Has that helped?

Neal: Yes, we have a staff meeting every Wednesday morning and we talk about everything we did the week before. We talk about the good things and what we need to improve on. Before we were doing your program we only had staff meetings every two weeks. It's been great to refine those communications skills and take weekly breaks together.

Mullins: That's a good way of looking at it, weekly breaks. With you personally, what would you say to another business that has someone like you handling the phone, maybe no experience, but now the owner has decided to do a program like yours, to push the envelope with selling? What would you suggest that they do? I guess a better question is, what did you struggle with when the idea is that everything is about sales and pausing and listening?

Neal: To be honest, what was difficult for me in the beginning was that I was expected to record my calls for training and I've never had to do that before. It's very uncomfortable.

Mullins: I do recall that you were upset during your interview.

Neal: It's not comfortable to be put in a situation where you're going to be criticized. However, having gone through many bad phone calls critiqued by you and my whole team, I learned so many things from it, I almost want to always record my own calls so I can learn another good lesson. You know, a lot of tough questions came out of those calls and because I had them on tape I was able to remember, "Oh, I had a really tough time with that. Can you help me out with it?" and you were great with giving me suggestions to try this type of phrasing, but it ultimately made me way more comfortable on the phone and more equipped to deal with touch telephone conversations.

Mullins: You're giving advice, Emily, to that person at another business who's resisting being recorded, the change in the office, feeling like now they're not taking care of customers or clients because they're trying to sell.

Neal: It's understandable and that's how I felt in the beginning as well, but ultimately you realize that by bettering your own communication skills you are actually taking better care of the client in a much better way than you were before. A lot of the questions I had before I didn't have the answers to and I would stumble over my words. I didn't know how to convince them but in my heart I knew we really needed to see them and I didn't know how to get them into the office. It is sales, however, you're doing them a favor, you're helping them more by getting them in the office.

Mullins: It's a relief, don't you think, because you're not supposed to have the answers, the doctor does. You're supposed to wow them and make the appointment, right?

Neal: Yes.

Mullins: Was it really a tough beginning for you?

Neal: It was really uncomfortable because you play the calls back for the people in the office and we critique each call one by one and talk about it. Getting past that, I do like listening to the call a lot because you can't get tone of voice in a transcript, or the pauses, or the feel of the caller.

Mullins: Off the top of your head, what type of financial improvements have you seen in the business over the past two years?

Neal: I have seen a huge increase. We have a lot more consultations coming in with the very first step and once they come in we have a lot more people that are going forward with treatment because they have gotten in to meet the team and doctor and they really understand how we can help them. We went from getting four or five consultations a week to now getting 12 consultations (new patient appointments) each week.

Mullins: Your internal office communications also help with new patient appointments and with patients accepting treatment, wouldn't you say?

Neal: Yes, our communication has definitely improved as well. It all ties into the experience that the patient has in the office.

Mullins: Are there certain things that the practice can do as a group to improve new patient flow, but that also helps the primary person answering the phone get through the tough time of critiquing calls and the difficulties of change?

Neal: Absolutely, you need to be supportive and realize that everyone makes mistakes. We're in this together. While listening to the critique you have to remember what it's going to be like for your teammate. In our office now, it's more like, here's what we should have done differently during that call. You can't break people down.

Mullins: What helped you grow for the six years with Dr. McAnally? I mean, it's not just about the phone coaching and working with Chris Mullins, because the bottom line is there are so many important things that a business — whether a dental business, tree service, health insurance, automotive repair, real estate, attorney, pest control — needs to know about running their business so they don't have team members that leave. What kept you there?

Neal: Lots of different things, our practice helps people. It was satisfying to know that we as a group are helping others. When our communication improved it was very supportive.

Mullins: So everything comes back to team communication, recognizing the successes and lessons in such a positive way that you'll want to come back and learn more, is that right?

Neal: Definitely.

Mullins: What about bonuses and reward programs?

Neal: If you turn 10 inbound phone calls into a scheduled consultation (that show up) within a three-week time period, there was a bonus of $250. Then we were making followup calls, so on outbound calls, if we scheduled 10 consultations in three weeks there was a $500 bonus.

Mullins: As a team or individual?

Neal: Those are individual goals, we also had team goals for the quarter, so if we changed our consultations and converted them into full treatment, and if we met our quarterly financial goals, we as the whole team would get a percentage of that.

Mullins: The magic number was that as an individual 10 inbound appointments over three weeks got you $250 and if you got eight or nine, you didn't get anything?

Neal: Correct, but it's funny, when you get to that eight, it really makes you determined to turn two more calls.

Mullins: Does it really?

Neal: Absolutely, we're racing for the phone saying, "I got it!"

Mullins: [laughing] No way! Awesome! Did you keep track of that daily or weekly to keep someone's name on it?

Neal: We looked at the totals every two weeks because that's when we processed payroll and everyone could see who was getting a bonus.

Mullins: Did you keep a chart?

Neal: We kept a chart on the wall to show progress throughout the year of the total of the bonuses.

Mullins: What about a daily target for the number of appointments?

Neal: Each person had their own tracking sheet of how many calls they had turned into a consultation (appointment), so we would look at the tracking sheets every week to see how we were doing. "You've got nine consultations, you need one more!"

Mullins: So, if you had 10 consultations a day over three weeks, did you get the bonus?

Neal: No.

Mullins: How often did you hit it?

Neal: I hit it often because I am the primary phone person, but it really influenced others to help with the calls rather than having people standing around saying, "Who's going to answer the phone?" It quickly changed to, "I'll get it!"

Mullins: Was anyone upset, like, "Hey, that's not fair that we have the same goal because we don't answer the phone as much"?

Neal: Not really because everyone has the opportunity to be near the phone — it rings all different times of the day.

Mullins: Did the rest of the team ever win bonuses while you were there?

Neal: Oh, yes.

Mullins: Did the rewards and bonuses help? Do you think you still would have improved sales if you didn't have them?

Neal: Yes, I would have because we still had long-term goals, but it made a big difference having the individual rewards and goals.

Mullins: Now what do you say to someone that still feels they have to give a price over the phone?

Neal: You know, honestly, you can't give price because you have no idea what the situation is. The easiest thing I have found is to deflect to the doctor because I'm not the expert, Dr. McAnally is. That lesson was huge for me because it wasn't me telling the patient that I wouldn't give them that information, rather it was me saying, "Gosh, I wish I could, but I have no idea because I'm not the expert with that, so the very best thing would be for you to come in and speak with Dr. McAnally because he can tell you that."

Mullins: Let's say they called for teeth whitening.

Neal: We do some pretty specialized teeth whitening in our office, so again, I would direct them the same way because there could be a lot involved.

Mullins: Is there any service or product that you think you could have given a price for?

Neal: I really try not to because a lot of times people call thinking they know what they need, but when they come in we find that they need more than what they thought was involved.

Mullins: How do you handle insurance?

Neal: I let them know that we will submit their claims for them but usually insurance doesn't cover very much, so we have a lot of great financial options to help them spread out their payments to make their treatments more affordable.

Mullins: Do you have anything to say to folks that are looking to sharpen their axe with answering the phone and booking appointments and for the folks that might be having difficulty with the program and mindset of sales?

Neal: Just try to remember that the phone call is usually the first point of contact that anyone has with the office and that first impression makes a huge difference. So if you speak with confidence and let the customer know that they called the right place and they need to come into the office, you'll be fine.

Mullins: Tell me about the sales drills meetings that you have every week.

Neal: We go over the exercises you gave us on PAS (Problem, Agitate, Solution), learn to accept constructive criticism, offer empathy to someone that is being critiqued, focus on staying on task in the meeting, review the past week and what's coming up including the language with certain patients. Before your program we used to talk about the patient if they stood out, but now we talk about every single one that came through our office in the last week and every single consultation that's coming up in the next week. We really focus on why did they call us, PAS, what's their problem, what's going to be their motivating factor and how can we help them to see that we're the best place for them to be.

Mullins: What about scripts? How did you feel about them?

Neal: Scripts were very helpful because you don't speak from the script like you're a robot, but when your mind goes blank and you cannot think of what to say to the patient, you go back to the script and it will help you get back on track. It gives you a good baseline to go from, but you want to make it sound natural. Scripts give me the answers to the questions that I know are coming.

Mullins: How did you organize them?

Neal: We created a small binder full of the various scripts. We would review them once a day and if we knew we were going to have an ad coming in, we reviewed for that. Having the binder in front of me was common.

Mullins: What about your outbound calls?

Neal: They weren't the crowd favorite, but we did them and had scripts for them and we have great incentives for them.

Mullins: What were some of the reasons for outbound calls?

Neal: A couple were if they booked an initial consultation, but needed to reschedule. Or, they just want information sent to them; we would follow up one week later.

Mullins: Any final word, Emily?

Neal: You know, Chris, I really just appreciate the effort that you put into our office. I know it's not easy to have people kicking and screaming with a new program, but we really appreciate it and I definitely have seen a lot of improvement.

Mullins: I really have enjoyed working with you, Emily, and watching you grow. Good luck in your new roles.

AHA Notes:

Chapter *Seven*

Mystery Calling / Monitoring the Phone

"The bleeding always stops." – Anonymous surgeon

Businesses are bleeding... The way to stop the bleeding is to clamp it off. Here's how we go about it.

Stop the Bleeding!

As consumers grow ever more disgruntled with customer service, America's retail stores, restaurants and airlines are investing serious time and money to keep their employees on their toes. Their secret weapon: instant feedback from an army of undercover shoppers.

The best way to avoid call reluctance and lost customer sales opportunities is to monitor your staff daily, record their calls and play them back for all to learn. Of course, you also need to give them individual feedback. As always, check the laws in your state regarding recording calls.

How to Break the News to Your Staff

You're probably wondering what that's all about. I'll tell you shortly, but first, here's how you can make the most of my advice.

I want to remind you about the art of translation. The idea is for you to be in the habit, have the behavior, of asking yourself every time you enter an educational moment, read a book, attend a class, listen to a CD, go to www.GreatBottomLine.com for weekly Off-the-Cuff teleclinics and even observe others, how can I take what I've just read, heard, observed and apply it to my own situation? I guarantee that you can if you're adept at translation.

I suggest you have pad of paper and pen ready to go. Draw a line down the middle of your page. At the top of one of the columns write in big letters the words SIDE NOTES. This is a strategy I teach students whenever they're in a learning moment. You know, when you're in a seminar or reading a book, attending a meeting or even listening to CDs and suddenly you hear something that excites you. Those are what I call side notes. Write down those items that excite you because those are probably the actions you'll be able to execute right away.

Let's assume for simplicity's sake we all have the right people in place.

You need to understand that how your team reacts to "breaking the news" regarding how they handle the telephone is critically important to the success of your message to them. Your mindset is key — you, the owner, and your office manager.

Here's what I'm finding on a regular basis when working with clients: The business owner doesn't have a consistent system in place for how to use the phone for inbound calls.

Or, they had a system, but they don't follow it, or they had weekly sales drills but got away from it, or had incentives but not any more, or they had a phone script but don't use it anymore — you get the idea.

It's up to you. Your team will follow you, you're the one that they want to impress the most.

They also watch you and follow you. So, if you're burnt out, they're burnt out; if you're sloppy they are; if you're not systemized they're not; if you're late or making excuses, they know and will do the same! Resentment is eminent.

And, if your attitude is, "Hey, I'm paying the salary I can do what I want," they'll get that message, too. The message again is, it starts with you and your mindset, how you feel about yourself, your business and your future wealth. And it all connects to the telephone!

You must make your phone, your sales efforts and your customer service (internally and externally) extremely important. Systemize and stick to it. Your team will follow. I guarantee it!

It takes time; it's a building process. Creating new behaviors is about making decisions, then dumping old habits to make room for new ones. Everything you do in your business with your staff must be from a positive standpoint.

The educational portion of your business, the ongoing coaching and training you're providing for your staff has got to be based on relationships, communication and sales.

So, when you tell your folks that you'll be doing mystery calling or "checkup calls" as we call them in our phone sales program, they won't like it. They'll initially be concerned, afraid, threatened or feel like you don't trust them and that you want to catch them doing something wrong so you can get rid of them or lecture them. (Be sure to check the laws in your state regarding recording calls.)

This is normal and you should expect it. However, if you've been focusing on your own mindset with your business in all the areas I mentioned earlier, it will be easier for them because of the success culture you've created. They'll still be uncomfortable, but it will be a little easier.

Resource: Go to www.GreatBottomLine.com to learn more about our easy-to-use Call Tracking System.

Don't back off or get nervous, remember, this is your business, you make the decisions. You've got to monitor your staff daily, you've got to track your incoming calls with the telephone. The telephone is the single most important instrument in your business. Without the phone and good staff, in that order, you've got nothing! No matter how good you are, you're done.

Ease into this process, explain to your staff that this is a positive program to help them be sharper than ever, to do even better than they already are.

Notice what I just said, better than they already are doing. That's a better message than, "we want to fix what you're doing wrong" or "we want to correct your mistakes."

Now that you've been having checkup calls done, you want to share the news with your team. So...

1) Calm down

2) Breath and relax

3) Remember your purpose (which is to be able to influence your team to repeat the great things they already do and quickly tweak the areas that need improvement. Notice the language I'm using.)

4) Have a plan before you talk to them about their critiques. Take the time to listen to the call recordings yourself and make notes about the good ones first and why it's important for them to continue those good points. Then share. Many folks don't even realize the good they're doing while talking to customers; it's part of their routine, so point it out, that way they can do more of it.

5) Next, share the specific areas that need fixing – give specific examples.

6) Share why it needs to be fixed.

7) Make sure that you also offer the solution, how to fix what needs fixing.

8) Make this time fun – smile – you want your team to look forward to your critiques and even ask for them. Believe me they will, this has been my experience.

9) Hold your team accountable for using their phone script or phone slip, which you need to make sure you have in place. We have our own 10-point script outline that our clients use.

10) Teach your team how to use the script, practice with and reward them.

11) Before you talk with them, you need to know what you want to fix. Put together a lesson plan for how they can fix it, include how you'll help them and when you'll meet again.

12) Commit to your weekly sales drills, listen to calls every week as a group critique together — no excuses.

13) No one likes to hear their own voice being recorded. I personally don't like to listen to my voice, but I do it so I can always be focused on improving. When I was doing public speaking, I was constantly recording my business talks and playing them back.

14) Remember, be sensitive. You're dealing with people, fears, rejection and self-confidence. You need to work at building them up, not tearing them down.

15) Systemize this process.

16) Teach your team to self-critique.

17) Pay attention to them, guide them, coach them, encourage them to visit with you, watch your tone, your words, your body language. Remember, you're always being observed.

18) Make it important, which means you won't drop the ball. Keep at it, because they're waiting for you to drop the ball since it's likely you have done so in the past with great ideas that have gotten you fired up. Not this time though. First you need to build the trust with your staff, even if you've been at this for 20 years. Second, your telephone and your staff are critically important to you, without them you have no business!

19) Give your staff the tools: scripts, outlines, FAQs, objections lists with the correct answers, role-play, provide coaching and support for them.

20) Peer recognition – encourage this, require it and systemize the process.

21) Catch them doing something right, daily.

22) If you're thinking, "Chris, they should know how to do it by now..." that's the wrong attitude. They're people, we need to be continually reminded of what to do and how to do it. And, you need to be sure that you've given them the right tools, guidance, information, training and ongoing coaching and mentoring. It's just as important for your staff, your business and your bottom line as it is for you!

23) I can guarantee you if you don't handle this moment, breaking the news the right way, sabotage will absolutely show its ugly head with many of your team members.

24) Remember to create an ongoing reward program and recognize your staff, each team member, for even for the smallest improvements.

25) Here's an example of how your introduction can go. It may sound something like this: "Great news! We're going to have an outside coach help you and me perfect our phone skills, sales and customer service to improve our bottom line, the number of new customer appointments we secure and deals we close!"

Your Business' Most Important Tool

Remember, our telephone is the one tool that opens or closes the door to the business. It must be done well.

Why? Nothing else matters if our phone is not answered well.

Handled well means prospective customers call to schedule and keep their appointments. Established customers keep their appointments.

The first voice that the customer hears impacts how we are thought of. If the gate is closed, no marketing will open it.

And if that gate is closed on day one, it is unlikely you can reopen it for the customer that just hung up without making an appointment or giving you their information. A closed gate means money is needlessly wasted on followup mailings and external marketing.

Have you ever seen a business burn through $10K a month in external marketing and have less to show for it than another business only investing $1,500 a month in its marketing just because of their phone? You have? Was that business your business?

What to Do With Our Phones?

Measure. That which we measure can improve. How do we measure?

1) Keep a phone log.
2) Self-record phone calls for critiques at team meetings for training and quality assurance. (Remember to check the laws regarding recording in your state.)
3) Review your recordings and critiques from your phone sales training program.
4) Talk about the recordings with the staff and how to improve each one.

The Phone Log at Its Basic

Use a plain spiral notebook; it need not be fancy. Write down every call. Record the date, time, name of caller, caller's contact information, staff member who took the call, result and dates for any followups.

At your staff meetings (weekly sales drill), look at the log to see what is happening and who is calling your business.

Use the log to identify questions and answers that the team must have at the ready. It can help identify scripts needed for review before answering the phone.

Some customers have such high fear that it will prevent them from showing up even after the appointment is made. We need to detect if a person is fearful and have the right script to handle these people.

Share stories of other customers in similar situations.

There are questions that occur in a customer's mind when they call. (There are also questions that a person might not tell you.) We have to answer the questions in a way that tells the customer, "you have chosen the right place, you can feel good about your decision. We have handled people in situations just like yours and you will receive the highest level of quality, care, skill and judgment." A customer needs to feel good about making the appointment and actually look forward to their visit.

The Self-recording of Calls for Training and Quality Assurance

At first staff will be resistant or timid to record their own phone calls. Once they understand that this really is to help them improve their abilities and to make their jobs easier, they will become very interested in bringing calls to your team meetings for self-critique. Once they understand that there will also be mystery call critiques performed, they will be far more likely to self-critique calls so that they get better between regular mystery critiques. Again, check with the laws in your state regarding the recording of phone calls.

Mystery Shopping! Customer Service IS Sales

I once read an article in *USA Today* about how hotels are trying to make your stay more personal. They hope you come back again and again.

I thought you might enjoy reading this, but I do have an assignment for you involving the art of translation, which many of you have heard me talk about.

Get your pad of paper and your highlighter out and make a list of the ideas you really like. Next to each one write how you can translate the same idea into your own business, then pick one and decide on a date to execute it.

With competition keen these days, managers are making it their business to get to know their guests on a personal basis. They know that special attention keeps them coming back. For instance:

- Hotel bartenders serving your favorite drink before you ask
- Special water, blue chips and blue-cheese dip that you like
- A soothing CD in your room after a very stressful day

- Handwritten notes welcoming you back
- How about as you rush in from a rain storm you find your favorite water with the note, "It turned a bad day into a good one."
- Housekeepers notice favorite toothpastes
- Doormen pay attention to when and where travelers go

All the information gets entered into your profile database system. Of course, strategies like this should be used for regular and infrequent guests to keep them coming back for more!

All of these ideas are great and will certainly add value to the experience your customers get, but you still must inspect what you expect even with ideas like this. How do you know it's being done the way it should be done? You don't unless you do on-site mystery shopping and mystery checkup calls. Mystery shopping is simply a quality control training tool, support for your team. That's the message you want to give them, otherwise it will be viewed as if you're trying to catch them doing something wrong.

It's Now Truth Time! Do the Math and Share It With Your Team

Here's how one client shares a lesson from one Phone Sales Program session with the team through an internal staff newsletter. You can do the same. However, I do recommend that you talk to your team in person about this first and follow up with internal newsletters proving the purpose over and over again. You don't have to outsource your mystery checkup calls and phone coaching program, you can do this yourself internally.

> The mystery shopper has been providing us with very good ideas as to how to be better at what we do and to separate ourselves from any and all competition. I know that I am much more careful about what I'm saying than I was before.
>
> I believe that it's now "truth" time. There is something that you need to know. An average direct mail piece costs 49.5 cents each. That is the cost for printing, shipping and postage. It is very common for us to send out 10,000 pieces per mailing drop. A very good response rate is .5%, or 50 calls. If we receive 50 calls from the mailer, you can correctly say that each call costs us $99. A more common response has often been 25 calls, or $198 each. Yes, sometimes the responses have been much lower, doubling our cost per appointment.

That means that if we mess up on just a few, our cost per appointment goes up dramatically. Think of $100 bills disappearing right in front of your eyes for every call that does not yield an appointment. I believe that you can now see why we're doing this mystery shopper program and why it's so important that we all work hard to make an appointment on every call.

I'll go a step further: TRACK your incoming calls — whatever it takes. If you have the opportunity to do this automatically through software with no manual labor, great! If not, pen and paper work just fine for instilling the sales mindset in each of the folks that work on your team, especially those that answer the telephone. Whatever you do, just begin!

Track the number of calls you get daily and follow the action that is taken for every single call. Of course, you also want to have a tracking system in place for the number of appointments you close each day.

Marketing + Incoming Calls + Ongoing Phone Sales Training + Ongoing Monitoring Calls + Daily Goals + Tracking Incoming Calls + Sales Talk (talking about the numbers) + Recognition = APPOINTMENTS = SALES.

Resource: To get more information on automatic messages, call tracking and more, go to www.GreatBottomLine.com.

12 Big Things That Go Wrong With the Front Desk Phone Team

1) Calls are too long call — time yourself to measure how long you're talking on each call. Focus on shortening each new call, otherwise you'll end up becoming friends and nothing more than a good conversationalist, not a sales guy, which equals zero sales.
2) Always ask permission to speak when you call a prospect, no exceptions.
3) Dig deep — get inside the prospects' head. What prompted you to call today?
4) Practice PAUSING with each new call.
5) Interrupting and talking over prospects. Stop focusing on getting through your agenda for each call, which is a sign that you need to reexamine your purpose. Your purpose should be to close as many appointments as possible each day. You can't find out what your prospects want and need if you're

interrupting them with your own thoughts.

6) You need to find out what it is about your offer that they don't like before you make multiple offers.
7) Listen to the prospects; what are they saying? Can you repeat back what they said in the exact words they used?
8) Pay close attention to the language and vocabulary of your prospect. This will help you to know what to do or say next and you'll certainly find out quickly how serious they are.
9) Practice slowing down your thoughts before you speak.
10) You're excited, which is understandable, but you're not allowing the prospect to finish sentences or thoughts at crucial discovery points of your conversation. Therefore, not only are you missing several buying signals, you're also not asking for the sale, the appointment.
11) Practice having a sense of urgency. Speed is critically important in all sales situations at every single point. If you have a sense of urgency in your voice, your prospect will.
12) Finally, you can do this, but first, you need to decide if you're in the game or not. But what you have to realize is that the telephone is the most time efficient, least expensive tool you have. It's just a tool, nothing more. Just because you're selling appointments doesn't mean you're less of a person or professional. Certainly it's not who you are.

Resource: Go to www.GreatBottomLine.com to learn more about how to implement the Cell Phone Technique to capture new customers.

AHA Notes:

Chapter Eight

Setting Up Your Own Internal Program

24 Secrets to Coach Your Team to Be Winners

1) WARNING: Stick to the phone script. Many times the front desk person can get comfortable in their own routine. They start to feel confident which is good, but for some it means they get lazy. They start to create their own phone slip (script) which means they lose focus on what works. Don't let this happen! The script that works is your best friend, influencing your prospective and established customers to buy — what they wanted to begin with or they wouldn't have called you. If you get lazy about this and allow your team to deviate from the script you run a very big risk of losing many appointment opportunities. Scripts work, so hold your feet to the fire.
2) Provide a list of common objections with the correct answer available.
3) Practice how to answer objections.
4) Use a script, test your script and then keep using the one that works the best.
5) Practice how to use the script.
6) Make sure that your team knows everything about your business before they get on the phone.
7) Create a testimonial cheat sheet that your team can quickly refer to. This helps to build their confidence, as well as sharing stories with prospective and established customers.
8) Coach your team on how to use testimonials; you can't just say, "Use this." Show them how they can weave different stories into their conversation.
9) Provide ongoing coaching, motivation and recognition. This is a must, because you're dealing with people. Coaching works for you and your team needs coaching, too.
10) Daily pop quizzes are a great way to keep your team focused, moving forward, motivated, sharp, excited and knowledgeable about your business.
11) Develop ongoing contests and bonuses.
12) Do the math. Track and talk about the sales numbers (appointments) daily with your team.
13) Teach your team how to focus on one call at time. Visualizing is a great tool for staying focused on the next call. Think about Mary, the next prospect in line to call. What does she look like, how can you be of service to Mary?
14) Voice message script. Create an outline for the messages your team will leave. Again, hold them accountable for using the script.

15) Provide headsets that work for your team. It's faster and easier for them to navigate and cuts down on fatigue.

16) Observe your team as they make calls. Listen carefully to the tone of voice. How are they doing, really? Watch their body language. Should they be in this position?

17) Teach your team to use good listening techniques by paying attention to the tone, the voice, the breathing and background noise of the customers.

18) Teach your team to keep detailed notes on each call, especially documenting the exact language and vocabulary of your prospects.

19) Stop treating your front desk team as low level employees! Without the telephone, you're out of business. Your inbound calls are critically important to your business. The inbound call gives the very first impression of your business. Pay close attention to how your phone is being answered and monitor those calls to be sure you're capturing all sales opportunities. Give your team all the important ongoing coaching, training and information they need to be successful. Set them up to win, not fail.

20) First, you must decide that the front desk position is critically important to your business. Roll out the red carpet! Stop dealing with turnover. Sell with integrity.

21) Monitor all team members that use the phone. There's no way around this, you must inspect what you expect, this is your business. However, you must go at this from a positive point of view. The purpose of monitoring calls is to listen to them to identify all the great things that are happening so you can repeat them and all the areas that need to be tweaked fast. However, it depends on how you view this strategy and then how you break the news to your staff about this new way of coaching. In fact, this technique is the fastest way you can grow your business and keep it growing.

22) Teach self-critiquing. Your team members should know how to critique their own calls and then provide you with feedback. No, this doesn't mean that you stop, unless you have this part of your training outsourced.

23) Implement a bonus plan (commission works best) that promotes MORE business which will make even more money for your front desk and all team members that touch the phone.

24) Have fun! Don't give up! You can do this!

Resource: Go to www.UniversityOfPhoneProfessionals.com to learn about on-line training and pop quizzes.

Know Your Numbers! Mullins' Tour de France (not!)

I'm getting a baseline for my cycling. Wow! I'm so excited, I've been searching and searching and have finally found a personal trainer for cycling. His name is Dan and he and I meet twice a week.

Our first meeting was basic fitness information, measuring my bike, checking my bike out, asking health questions and taking my measurements. Next, we'll be doing a three- and five-minute time trial to get a baseline of my current performance — sound familiar?

My long-term goal is to complete a 100-mile race. That's a 12-week training program and I can't wait! My short term goals are different each week.

More on Chris' Personal Cycling Trainer and How It Relates to Your Sales Business

This guy is really big on the numbers. The numbers tell all; they quantify everything that I do on my bicycle.

The first day, I had a three-mile time trial (a 5K) on my indoor trainer. I know I already mentioned this to you, but I felt like Lance Armstrong; you know that face, that serious, focused, disciplined look.

I'm not sure why, but when I'm on my indoor trainer, I am much more focused when I look down at the floor. Maybe it's because that's when I'm in that pushing zone, the sweet spot. It isn't easy, but I really love pushing myself.

Dan had his stopwatch and was asking me where my level of exertion performance was at: hard, very hard or extremely difficult. The idea for three miles was to go as fast as I could while pacing myself to finish.

As I mentioned earlier, Dan is trying to get a baseline for my current performance level so he knows what my weekly goals need to be to get to my final ultimate goal — the big race.

Next, he'll create a customized training schedule that's made up of cycling, and because I enjoy running, he'll include that along with the three-mile runs I already have scheduled.

Dan can now create weekly goals for me with a new higher goal each week all to get me to that century ride in the best possible shape. Each week we'll ride together and monitor my progress.

In one month we'll do another three-mile time trial to see how I have improved.

SALES LESSON Know your numbers. Know what gets you excited and motivated to push each day to be a top-performing person, not just in sales and business, but in life. It's up to you — not your family, your friends or your boss. I was asked by a client, "How do I motivate my team?" My response is you really can't, they have to be self-motivated. Of course, you need a system in place for hiring correctly, but setting that aside, there are some things that you can do to initiate and foster a healthy, motivating, reward-team, focused environment. For starters, get to know your team individually (all of them). Ask what motivates them. Ask them to really give it some thought and get back to you.

Hint: It's not money. Tell them they can't come back to you with the old standby; a raise or money. To help them figure it out, share what really motivates you. As always, have a list of questions, thoughts and ideas in front of you. This means you need to *prepare* before you schedule time to talk to them.

POWER NUGGET™

Know your numbers... know what gets you excited and motivated to push each day

Get in the habit of catching people doing things right. You personally may feel like, "For crying out loud, we're all adults. Why do I have to tell them they did a good job? Why do I have to keep training them? Shouldn't they know all of this by now?" No. One thing you have to understand as the business owner — the leader, the manager — is that you're dealing with people; we're all human, we make mistakes, we have another life besides work. We have baggage, bad habits and self-defeating behaviors. As long as you decide to continue working with people, you need to become more educated on how to coach and work with them. You're in the lifelong learning and teaching business.

Put Your Entire Team Through Role-playing Based On PAS (Problem, Agitate, Solution)

A top salesperson not only knows WHAT sales is — how to use and apply PAS, close, follow up, build relationships, time management, multi-task, knows and understands the power of words — he or she must also have the voice (tone) and know how to use it during all points of communication with each and every prospect and member.

How to Become a Master Closer With PAS

I'm sure many of you have seen this before, but the question is do you actually use it in your business?

Here's a great trial close example:

If I show you how you __________________ would you be interested in knowing more?

Before you can trial close, you've got to know why they're calling you today. Get inside their head. Ask questions and *listen*.

Try the PAS Formula

Many of you have probably already heard of PAS. With this formula, you first state a problem and secure the prospect's agreement that the problem exists. For example, you might agree that it's a terrible feeling to be without teeth. Or, you might agree that it can be intimidating or scary to have to look for a new doctor. [PAUSE]

Next, you get the prospect agitated about the problem, perhaps realizing that not having a mouth full of healthy teeth is very dangerous. For example: "One of my newest customers had a terrible experience. She had to have all of her teeth replaced because she had severe dental problems due to her diseased mouth, making her a strong candidate for frequent heart attacks, strokes and even in jeopardy of dying sooner." [PAUSE]

Finally, you produce the solution. "Wouldn't it be wonderful if there was an alternative to protect you from those daily surprises?" [PAUSE]

Try the team exercise on the page to the right.

Resource: Go to www.GreatBottomLine.com on how to get *Chris Mullins Nuggets for Sales®* print newsletter and read about how to do the one-on-one weekly staff meetings.

Team Exercise: Try this team exercise in your next weekly sales drill.

The problem: ______________________________

Agitation: ______________________________

Solution: ______________________________

The problem: ______________________________

Agitation: ______________________________

Solution: ______________________________

The problem: ______________________________

Agitation: ______________________________

Solution: ______________________________

The problem: ______________________________

Agitation: ______________________________

Solution: ______________________________

The problem: ______________________________

Agitation: ______________________________

Solution: ______________________________

The problem: ______________________________

Agitation: ______________________________

Solution: ______________________________

The problem: ______________________________

Agitation: ______________________________

Solution: ______________________________

AHA Notes:

Chapter *Nine*

Sabotage

What a topic! Sabotage, unfortunately, is everywhere, in everyone's business, but many of you already know that and choose to look the other way.

Some signs of sabotage with training programs, especially mystery calling, can include comments like...

1) We don't get questions like that, and even convince you, the owner, that well, we don't get calls like that, so the program must not be a good one or we don't need this training.
2) I knew it was the mystery caller, so why should I bother trying? I'm too busy for this stuff.
3) Our sales are good, why do we need this?
4) I'm not a salesperson, I won't do it.
5) I won't use a script, I don't believe in them.
6) I've been here for 10 years, I know what I'm doing.

This Is About Teaching — It's Training! Have Fun!

The whole purpose of training, critiques, monitoring and coaching is you want to impress to your folks that this is *training*. It doesn't matter if a call seems real or not, the idea is for the front desk expert to be on top and ready for any type of call, to be sharp and always in a state of readiness to *ask for the appointment*, to test themselves to see if they can have that automatic behavior and habit. Can they be thrown off guard and still be on track?

The idea is really to get in the sales mindset in all interactions with established customers and prospects by phone and in person. Also, to become such an expert that you see opportunity in all moments and that you can't be distracted at any time, in person or over the phone with a question that might seem different or not normal.

With this training you're being tested with our Phone Sales Program™, so you can use that training, instinct, habit and behavior with all calls. Just follow the script and ask for the appointment. Get through the script and practice using it internally with team members. Have fun, learn and grow. This isn't supposed to be easy.

As we move forward in our program, some questions will be repeated — the ones you tend to get — but we'll also be making the questions more difficult. We're trying to turn front desk experts into sales experts.

As long as you and your team look at your monthly Score Cards™, listen to the calls, attend classes monthly and/or listen to the replay, and as long as you talk about the program progress and how to keep on improving, you will all get there.

Employed and Unhappy

Did you know that people don't go to work merely to perform a function? They do rather to have...

1) A sense of accomplishment
2) Self-esteem
3) Sanity
4) Community

Staff Becomes Unhappy Because

1) Bosses don't know them or care
2) They don't know why their job matters
3) They have no way of measuring the good effect of their job on others

Sabotage

1) Those that resist the change, the you-make-a-difference theory. They say something like... "This ain't rocket science and it sure ain't the center of my life." So, they get fired (if you dare)!
2) Others think it's too touchy-feely for the workplace — "too much love in this room" cheering each other on. You can laugh along but don't back down.

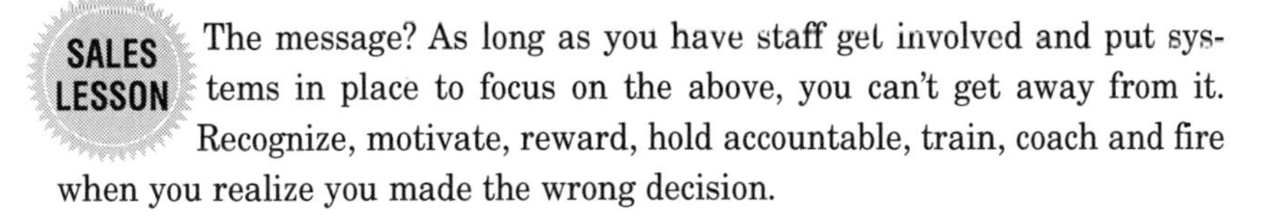

The message? As long as you have staff get involved and put systems in place to focus on the above, you can't get away from it. Recognize, motivate, reward, hold accountable, train, coach and fire when you realize you made the wrong decision.

> **Resource:** Go to www.GreatBottomLine.com to hear an embezzlement background check interview with Kevin Connell of Accu-Screen Inc. (www.accuscreen.com).

New Customer Call Horror Stories

When you call audit your own office telephone calls — you do call audit your own office on a regular basis, don't you? Good. Have you ever heard something like this?

1) "I'm on lunch right now, can you call back?"
2) "Truthfully, you can get the same _________ done at XYZ for a lot less."
3) "I think it's very expensive, but if that's what you want, go ahead."
4) "Okay, if you don't want to make an appointment now, we're not going anywhere so call us when you're ready."
5) "All our services come right down to money."
6) "You can compare that _________ to buying a car; do you want a KIA or a Cadillac? It depends on what you want to spend."
7) "Yes, I think _________ are an expensive way to go. We're actually more on the expensive side, in fact, much more than most, but you didn't hear that from me."
8) "I know what you mean, I hate going to the owner."
9) "I'm helping someone else right now, can you call back?"
10) This call was a transfer to another person in the office... "I have a beast of a woman on hold, good luck."

I'd say these staff members are putting an awful lot of faith in the hold button, wouldn't you? How bad does it have to get before you finally decide to do something about it?

Horror Story: Pizza Shop Piles on the Cheese!

Here's a very good (and true) example of why you must have systems, monitoring and training in place and another good example of why you must use a script, guide, outline or cheat sheet with your inbound and outbound phone calls. Even though this story is about an in-person counter sale at a pizza shop, the team member has no guide, no "here's how we do things" and no supervision, guidance or monitoring.

Your team will make up the rules as they go along and it's not their fault. It's yours because you're not providing them with ongoing support, training or recognition for the steps they take along the way to get to the goal guiding them on how to fix what they might have done that you didn't want them to do. This is a good recipe for disaster for going out of business by having a revolving door for staff with poor morale.

Read On

One night I went to pick up my pizza. I have a pizza punch card — you know, so many punches, you get one free. This card had some spots for punching when you get a small pizza. I got a large, so I asked the girl to punch my card and she did; she punched two smalls. Now, I've gone there before and that never happened so I asked, "Oh, is that what's supposed to happen each time we get a large pizza?"

She replied, "No, I do it because it just doesn't seem right to do it the way they want."

What this person is telling you is that they don't support what the owner wants; her own personal beliefs of what's right and wrong are getting in the way, so she's making up her own rules. Remember, you're not your own customer!

This is a common problem in all businesses from the front office to the back office. Don't think for a minute that your team isn't making up their own rules — tons of them — day in and day out. You must fill the holes in your bucket.

For those businesses that aren't watching, aren't looking, aren't checking up on their team or looking over their shoulder, they're losing buckets of money and lost sales opportunities. Remember, what gets measured gets done!

> We'd love to hear your horror stories! E-mail assistant@mullinsmedia group.com and put HORROR STORIES in the subject line. *I dare you!*

Warning!

If you have team members that are telling you this, then you'd better run or get out the smelling salts because it gets a lot worse!

Team member: "I've been here for over 30 years. I understand the telephone and communications business. I can tell when the caller doesn't want me to ask their name or contact information or doesn't really want to make an appointment. They just had questions."

Internal Customer Service:
There's a "Cancer" Growing in Your Office This Very Minute

Keep this very important thing in mind while reading: How we treat our internal customers is ultimately how we treat our external customers.

Throughout my years in business, I've uncovered a disease more deadly to business than a stock market crash or bankruptcy. This disease spreads rapidly and can destroy your business with the blink of an eye. It's a disease that many don't notice or recognize even though the warning signs are plentiful. It's a disease that can be best described as "destructive behavior," allowed to exist and grow by management. This disease doesn't discriminate; it quickly touches all departments within businesses offering any products and services. Read, take notes, get ready and be open-minded, because unlike some diseases, this one has a cure.

During past training and consulting contracts, I've dealt with many types of businesses and people, both "virtually" and in person. More times than not the first issue that comes up with a client is getting along with others in various departments. I may be hired to give advice on how to increase sales; therefore I would spend time with frontline salespeople. During one particular class, I spent more time listening to complaints from salespeople about other team members and managers. It just so happens that this particular business is suffering from sales growth.

POWER NUGGET™

If someone has enough courage to bring a problem to a manager's attention, you'd better listen

Here's what's happening: people are thinking of quitting, people call in sick, take more time off than usual, sales are suffering, feelings are being hurt, people's baggage is being brought to the surface and the paying customer's needs aren't getting met to the fullest.

Most of the time when this sort of combustion exists in a company, the frontline managers will hear about it. However, I've found that most managers just chalk it up to normal complaints — they'll get over it. Wrong! If someone has enough courage to bring a problem to a manager's attention, you'd better listen. I'm not saying it's always 100% accurate; I'm not saying the person venting to you has no responsibility. I am saying that for someone to take the time to speak with you, something is brewing.

Now here's the part that you — as a manager and a company — have to be open-minded to and accept. There's a common MO that I've seen in these types of companies. See if yours fits the mold.

1) Lack of company communication as a whole.
2) Management doesn't involve employees in the decision-making process of new products, services, benefits, policies, etc.
3) Too many layers of supervisors, managers and directors.
4) No department and company meetings with good, honest ideas and feedback.
5) Overworking employees, not paying them for their time and pressuring them to do it all because they don't pay over 40 hours, knowing very well the team is working more than 40 hours.
6) Micromanaging. No employee involvement.
7) Sharing last-minute decisions regarding mergers, buyouts, sellouts, reorganization of offices having the attitude that "hey, this is business, do what you're supposed to do or we can just get someone else to fill those shoes."
8) No training, just throwing them in. No respect.
9) Giving supervisory responsibilities to people just to cover an area that you, the leader, should be taking care of.
10) Worst yet, not training these supervisors on how to be managers and leaders. This is probably the worst area for internal customer service issues.

The List Goes On...

Your company may offer the best pay, benefits, perks and office environment. Don't get complacent. This isn't all that keeps people. Employees want to know that leaders are listening to them, guiding them, helping them grow, care about them as people, know a little about their personal life and members of the team. This means that you're a manager, psychologist, mentor, teacher, detective and financial planner.

All people have feelings. You must think before you speak. If, unfortunately, the little buggers come out so quickly that you can't stop them, think about what you said. Make it right, apologize and accept responsibility. Some time ago I attended a meeting for speakers, and while sitting in the audience I heard a fellow speaker say... "You have the power to affect someone's self-esteem positively or negatively."

Think about it for a moment; the responsibility we all have with our attitude and our words.

The tonic for this cancer? This time-tested remedy is as old as my grandparents. Before I give it to you, I'd like to say that it's the same for managers and leaders, because in most case studies it's been the managers that have added fuel to the fire by not having their own appropriate business training.

The Remedy

Managers must pay attention to the team. Get to know them. Learn from them. Let them vent — to a point. Teach them your mission for the department, which should be aligned with the business as a whole. Find out what their personal and professional goals are.

Get to know them as individuals; their strengths and weaknesses. Provide training for the weaknesses, whether personal or professional, promote life-long learning, put them in jobs that will compliment their strengths.

It's much more cost effective to invest time, energy and money with your internal customers so they will always take care of the external customers. Why keep on spinning your wheels trying to find, hire and retain people all over again? Care about what you have, teach respect, ethics and compassion.

AHA Notes:

"The average business turns over 70% of their staff every three years!"

What happens when you lose a staff member who's been answering the phone?

You and I know that other business offices are trying to recruit your staff pretty regularly. One of the important pieces you'll be able to uncover while monitoring your staff is, is there a morale issue? Who can be fixed? Who's the culprit?

We don't want you to have to start over, but we also don't want to waste your time trying to fix people that just can't be fixed. You'll be shocked at the other bits and pieces of information we'll be able to give you about your business, the stuff that you thought just didn't happen at your office. It's time to stop the competition from taking your good people.

Do you think of your receptionist as just a receptionist? If so, that's what you'll get.

Have you ever witnessed your receptionist turn a customer away that was ready to buy your basic services? Then, when you approached her did she tell you that, "I know you only like to see the big stuff!"

You see, it's not your staff's fault. They really don't know any better. This happens because you think of the receptionist as just a receptionist, probably not with much organized focused training and coaching on business, sales, customer service and your goals and focus for the business.

The way your phones are answered is where it starts. It's the most important aspect to your business. You've got to coach your staff on new customer acquisitions, a brain cleansing if you will, phone rings = new customer! They must assume the close, it's that simple. It's up to your staff to control the conversation and make the appointment.

Business owners, you've got to put just as much time and effort in your recruiting as you do your marketing, and as much time with your phone training as you do your marketing.

You're in the development of people business. Research back in 1999 showed that as we move into the twenty-first century, people will stay with organizations where they have a chance to grow. They've identified these organizations as learning organizations.

> **Resource:** Go to www.GreatBottomLine.com to listen to a free interview on our Automatic Hiring Systems.

Interviewing Tip

Twenty percent talk time for the interviewer and eighty percent for the interviewee. The only way to find out what's inside your sale reps and/or prospective sales rep's head is to get them to talk.

When you, the owner/manager, doesn't understand that it is your responsibility to create a motivating environment, you create a void where any bad influence or negative thoughts (we have a lousy compensation plan; we have lousy services and products) gets in. Once that happens, especially in a sales environment (which your business is), you've lost control and you'll be dealing with all sorts of challenges like poor closure ratios, missed sales targets, team members that go from stardom to the bottom of the barrel, excuses about presentation skills, leads, appointments, turnover and days off. Remember, your job is to create a motivating environment that will influence sales to improve your bottom line.

Calculating the "Real" Cost of Employee Turnover

To calculate the real cost of employee turnover, you need to consider numerous areas affected by turnover. Top and bottom lines are affected and the recovery process is painful. The examples listed below will give you specific details on each aspect of calculating the real costs of employee turnover in any business of any size.

Advertising

Each time you replace a person you must advertise the availability of the job. Newspaper advertising, chamber, web sites, magazines, e-mail newsletters, association newsletters, direct mail, job fairs, booths, open house, marathon interviewing, etc. Then of course you must calculate the amount of time that's spent to create the ad and the advertising strategy. Once you calculate the cost of your time, don't forget to add in the cost of the actual advertising. To get your final costs take the salary of the individual or individuals that are responsible for this portion of the recruiting effort, track the hours spent, figure salary per hour and you'll have your time costs. Then, add in the actual expense of advertising. That's the real cost of the advertising portion of turnover!

Interviewing Time

Each time you interview a candidate, whether it's formal or informal, on the phone or in person, it's costing you money. You're spending money when you go to get the mail, when you open the resume mail, when you read the resumes, respond to inquiries, check references and discuss the position with internal team members. Once again, look at the salary for the individual(s) involved in this portion of recruiting and track the hours spent. That will give you your time costs. In addition, you may incur interviewing expenses to promote the position and/or your company. For example: you may take applicants to lunch or you may have additional managers in the company conduct the interviews so you can get different opinions. Don't forget to calculate the time spent by all that participate.

Training

Each time you train a new person it costs you money. Here's how: The person that does the hiring usually does some training, then asks a fellow employee to help out with training. The time you spend training, no matter how formal or informal, the time you spend explaining the job responsibilities to the fellow employee and the amount of time that they spend actually training the new person are all costs! Look at all the salaries involved, create an average hourly wage, calculate in the hours spent on this portion of the recruitment process and you'll have another piece to the real cost of turnover! Wait, it's not over! The costs of training are not just measured in time.

Lost Time for the Rest of the Team

Each time YOU, the leader, take time to focus on the recruitment process you take valuable time away from the rest of your team. You forget them. You're not focusing on their successes or challenges. Your business suffers, you lose track of current and future projects. You become overwhelmed, stressed, forgetful and perhaps even a little grouchy. You're the leader. You're the one they emulate, they watch your every move, they count on you and, whether you like it or not, they look up to you.

You see, in a quiet way your actions are giving them permission to act the same way when there's a change or challenge for them to deal with. They lose focus, organization is an issue, there's no motivation and attitudes are negative. And that's not all. When you have a leader that's not leading the team, you get a team that's not functioning as a team. They take it out on each other with pettiness and bad attitudes. The way they treat each other is the way they'll treat the paying customer. Your external customers are getting attitude, bad service, etc. That's your bottom line!

Effects on Employees

In addition to attitude being affected, workloads are affected for each member of the team. Each time you hire a new person you shift job responsibilities to different members of the team. Usually you don't even ask, you just pass it on and sometimes you don't even delegate to people that know how to handle the new responsibilities.

Of course, most team members want to do well and dive in, so even if they're out of their comfort zone they may not tell you. When workload increases, especially for inexperienced people, you create a plethora of problems. Employees will suddenly be out sick, take personal days or vacation time and you're so wrapped up in what you're doing, you may not notice. This isn't even the worse case scenario; other employees start to give their notice. They've just had it! The costs involved are infinite.

Take Action Now!

You can't be lazy about this. If you have a concern about a team member and how they're doing with their book of business — don't wait. Take action immediately! Remember, each individual team member is a salesperson and they can positively or negatively affect the bottom line.

It doesn't get better and it doesn't change, but you already know this. You've got to jump on it right away and say it like it is. Let's say you're concerned about sales and making goal.

1) Prepare (script) before you say something. Know exactly what you want to say and why.
2) Pick a day and time to have your one-on-one. If you've been reading *Chris Mullins' Nuggets® for Sales* (CMN) print newsletter, you'll be familiar with this one-on-one process. Or, for more information go to www.GreatBottomLine.com.
3) This should be a positive conversation. You really just want your team member to know sooner rather than later what's on your mind and what your expectations are. The idea really is how you can help them to quickly succeed.
4) Agree to what's next. How will you both move forward with a new understanding of where sales need to be?
5) This may include coaching, training, role-playing or perhaps just having the conversation will be enough. You may even find that you assumed that your rep would know exactly what you're thinking and exactly how you want business to be — not!

How to Hire Good Salespeople

Where are all the good salespeople is a question I'm often asked. Here's what my experience has shown to be of proven value.

Know What to Look for in a Good Salesperson

Evaluate the really good employees at your office. Pay close attention to your top performing people. What personality traits do they have, where do they come from, what are their parents like (large family, small family), their work history, etc.? If you do this on a regular basis you'll have an exact blueprint of the type of people you need to look for and hire.

Ask your top performers questions like, how did they get all these good habits, what's their secret? Read my monthly print newsletter, *Chris Mullins' Nuggets® for Sales*. Each month we do a personal interview with a Sales Hot Shot to find out what makes them so successful. Read each interview more than once with highlighter in hand and soon you'll have a great head start as to what to look for when interviewing candidates.

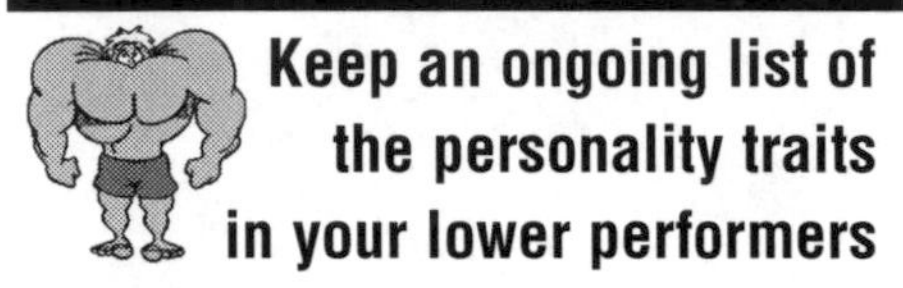

Birds of a feather flock together. Ask if they have friends like themselves who may be interested in sales work.

You'll also want to keep an ongoing list of the personality traits in your lower performers. Again, if you're dedicated to this, by documenting and tracking the facts you'll see a profile of the folks you should hire and the ones to stay away from.

Here's What My Clients and Colleagues Have Seen Over the Years

Most of the really good salespeople have grown up in a family business — parents who ran some sort of business while the kids worked there after school or in the summer. Most seem to have experienced the restaurant business as bartenders or wait staff — they are the best and you should build this into your advertising copy writing.

Other hard-working environments to recruit candidates include jobs where they would deal with the public and difficult or unhappy customers. Most of these types of jobs have demanding bosses — landscapers, cleaning companies, factory workers, retail — all with a proven history of working really hard. Folks who play sports are goal oriented and competitive.

Ask questions: What are your goals? Are you a team player? Great! Give me some examples. What's the last goal you achieved? How did you achieve it? Believe it or not, all of these questions are even more important than any level of sales success or experience in any industry.

I do suggest investigating personality tests to measure if they can sell, will they sell, will they sell here with these customers, products, in this environment. I used a predictive indexing tool several years ago while managing a team of salespeople with success. However, I've also had good results using my own blueprint method.

> **Resource:** Go to www.GreatBottomLine.com and click on Chris Mullins' Secret Bonus Hiring Question no one ever asks applicants.

Emergency Cheat Sheet for Business Owners and Sales Managers

Let's say you've just had to fire your receptionist or front-desk person. I suggest that you make a quick list of all the specific reasons why this person didn't work out and vow to not repeat any one of them. If that means you need to communicate differently with another person, like a supervisor, then do so.

My Big Thing

My big thing is guiding clients to NOT repeat the same. The list you come up with about the person you let go should include items such as personality, background, experience, interview, reference check, trained for sales, customer service, marketing/lead generation, telephone, business procedures, etc.

Be very careful not to treat the receptionist like just a receptionist, because that's exactly what you'll get! This person needs to be positioned as the expert at what they do: communicated to, respected and treated that way *internally* and that's exactly how they'll automatically position themselves *externally* with your clients and prospects.

You're basically creating your own blueprint of what NOT to do with staff to refer to often. In fact, I suggest you schedule time in your organizer to take a quick peek at this list one time each month.

AHA Notes:

Chapter *Eleven*

Self-critiquing and Role-playing

Self-critiquing, Role-playing, Sales Drills, Oh My!

Do this faithfully every week in your sales drill and your appointments will soar. Whether you have an outside expert monitoring your calls and teaching your staff how to use the telephone as a sales tool or not, you and your team must be dedicated to putting together your own internal training program. What gets measured gets done! This is a priority — no excuses! (Of course, check the laws in your state regarding the recording of calls.)

1) The front desk person should bring their digital recorder to the sales drill and play at least one call to be critiqued by the team.
2) Self-critiquing — the front desk person should listen to their own calls at least 3 times a day and fill out a check list of the areas they critiqued in each call (the script). So, if your script has 10 points, they should critique each point with each call, then pass the self-critique score sheet to the owner/manager.
3) Each team member should share what they feel was great with the call and why it should be repeated, but they must all share what didn't work and why.
4) Each person on your team that touches the phone should bring the phone slip (script) with them to the sales drill. In fact, each person that touches the phone for any reason, even to help prevent the call going to voice mail, gets to record their calls and get critiqued.
5) Do at least one role-play session in your weekly sales drill, because the sales drill is trying to stay focused at 45 minutes (ideally 30) per week. Once you get into the swing that will work, if you stay focused. Pick one scenario a week, everyone takes a turn playing parts, plan it in advance and make it as real as possible.
6) Track your incoming calls — all of them — and discuss this in your weekly sales drill.
7) PAS – discuss the PAS with each upcoming new customer that's scheduled to come in for an appointment.

Mystery Check-Up Calls

"Don't Slip Back to Old Bad Habits" Maintenance Program™

Chris' Five BIG Thoughts

1) Welcome them with, "Let me be the first to welcome you to ABC." This is brilliant because it's telling the caller that they've been accepted, it gets them to begin to think differently about the call. "Hmmm, I'm not just calling to ask my questions, to do my homework, I'm really calling to make an appointment," plus no other owner or other business does this and you stand out amongst the rest!
2) What prompted them to call today? You get to find out the real emotional pain... PAS (Problem, Agitate, Solution).
3) Ask for the appointment! No matter what. I know not all calls seem to be the right type of call for this, but my biggest point here is mostly for you and anyone else that takes calls to have the sales mindset and stay in automatic appointment mode. This is very important for behavior and habits. Keep in mind every single call you take is really training for the next call that comes after that.
4) Get contact information to send your free report or custom newsletter.
5) What about... How did you find us?

Raw, Unedited Example

[**RED FLAG:** One thing to remember is that when your phone rings it's because the caller, whether prospective customer or established, wants to talk a live person. They didn't say to themselves, "I want to call today because I want to talk to an answering machine or voice mail." They called because they want a live person and to do business with you. It's important to think of yourself as a consumer. How frustrating is it to you that when you finally get around to picking up the phone and dialing the business you get voice mail, answering machines or push 1 for X, push 2 for Y, push 3 for Z?

What's your average transaction size? You've lost it with answering machines. You've got to develop a system where you're not using a machine. Get an additional person to help with the phone — interns, part time, temps — even if to just take names and numbers. Have calls forwarded to a special cell phone number before and after office hours so that prospective customers and established customers will get a live person.]

Go to www.GreatBottomLine.com to learn more about this technique.

Chris' Five BIG Thoughts

1) Your voice must have a sense of urgency for all types of calls otherwise it won't seem so important to the customer and therefore selling appointments that show up won't happen.
2) Set daily appointment goals to sharpen your personal axe and performance goals, to compete with yourself from one day to the next and to avoid complacency.
3) Use a script or opportunity phone slip with each call. This will keep you focused on the call and the steps you must include to take the caller down the path of scheduling the appointment
4) Catch each other doing something right everyday.
5) Have fun! One call is just training for the next.

Chris' Eight BIG Thoughts

1) Focus on asking (selling) for the appointment to prevent lost sales opportunities. Bottom line: Y appointments is really what this is all about. You've spent thousands of dollars and huge amounts of time on marketing, testing and tweaking to finally get the right prospective customers to call and when they do, go for the appointment — quickly. That's really why they called, which is why they have a list of questions.
2) Always get the contact information to prevent lost sales opportunities.
3) No diagnosing over the phone, just sell the appointment. No diarrhea of the mouth, get them in to ask all their questions.
4) Your voice must have a sense of urgency for all types of calls otherwise it won't seem so important to the customer and therefore selling appointments that show up won't happen.
5) Set daily appointment goals to sharpen your personal axe and performance goals, to compete with yourself from one day to the next and to avoid complacency.
6) Use a script or opportunity phone slip with each call. This will keep you focused on the call and the steps you must include to take the caller down the path of scheduling the appointment.
7) Have fun! One call is just training for the next.
8) Catch each other doing something right everyday.

AHA Notes:

Chapter *Twelve*

WILs and Questions/Answers

What I Learned™ (WIL) Forms

This is a Phone Sales Program™ staff accountability tool to keep your staff focused on learning during each training session and to keep you the owner and office manager, focused on setting individual goals. In fact, I suggest that you have your team fill out WILs for your weekly sales drills and your weekly one-on-one goal sessions with them for the very same reason.

Here are various people's responses on their actual WIL sheets. They include not only what they learned, but what they will do specifically to apply what they've learned.

3 Action Steps Following Each Monthly Teleseminar

1. **Email your WIL to:** support4.chrismullins@bluetie.com
2. Or, you can Fax 1-866-694-2563 - **please "print" LARGE and clear**
3. Give a copy of your WIL to the Office Manager or Doctor to discuss your individual goals.

What I Learned (WIL) Cheat Sheet

WIL's are the perfect way to begin to hold yourself accountable *following* your 'monthly group teleseminar' with Chris Mullins. For optimum results stay focused on the specific action steps you'll apply. **<u>You do NOT need to use this form in order to submit your WIL; however you must answer the TWO questions below and provide ALL the information below.</u> ☺

It's important that you <u>discipline</u> yourself to do your WIL once you complete your call, otherwise the chances of it getting done and keeping all the important thoughts 'top of mind' are very slim.

** **<u>Give a copy of your WIL to the office manager or doctor to discuss your goals.</u>**

**Date June 13 **Name Aileen

**Practice Name ______ **Phone ______

**Mailing Address ______

1. What I learned during my call, <u>SPECIFICALLY</u>.

That repeating the problem back shows that you are listening to the caller, that you care & if you misunderstood that you are trying to understand.

2. How I'm going to <u>APPLY</u> what I learned, <u>SPECIFICALLY</u>.

I always repeat the problem to the patient to make sure I heard correctly and it also enables me to book the correct amount of time with the correct provider

Chris Mullins' Phone Sales Doctor™

3 Action Steps Following Each Monthly Teleseminar

1. **Email your WIL to:** support4.chrismullins@bluetie.com
2. Or, you can Fax 1-866-694-2563 - **please "print" LARGE and clear**
3. Give a copy of your WIL to the Office Manager or Doctor to discuss your individual goals.

What I Learned (WIL) Cheat Sheet

WIL's are the perfect way to begin to hold yourself accountable *following* your 'monthly group teleseminar' with Chris Mullins. For optimum results stay focused on the specific action steps you'll apply. **You do NOT need to use this form in order to submit your WIL; however you must answer the TWO questions below and provide ALL the information below. ☺

It's important that you discipline yourself to do your WIL once you complete your call, otherwise the chances of it getting done and keeping all the important thoughts 'top of mind' are very slim.

** Give a copy of your WIL to the office manager or doctor to discuss your goals.

**Date June 13 **Name DEBBIE M

**Practice Name________________ **Phone (   )________

**Mailing Address________________

1. What I learned during my call, SPECIFICALLY.

repeating the problem back to the patient helps make sure that you understood what they said and shows to the patient that you were paying attention to what they said.

2. How I'm going to APPLY what I learned, SPECIFICALLY.

When I am about to repeat the problem, I will say "Let me repeat it back to you, so I can make sure I got it right".

3 Action Steps Following Each Monthly Teleseminar

1. **Email your WIL to: support4.chrismullins@bluetie.com**
2. **Or, you can Fax 1-866-694-2563 - please "print" LARGE and clear**
3. **Give a copy of your WIL to the Office Manager or Doctor to discuss your individual goals.**

What I Learned (WIL) Cheat Sheet

WIL's are the perfect way to begin to hold yourself accountable *following* your 'monthly group teleseminar' with Chris Mullins. For optimum results stay focused on the specific action steps you'll apply. **You do NOT need to use this form in order to submit your WIL; however you must answer the TWO questions below and provide ALL the information below.** ☺

It's important that you discipline yourself to do your WIL once you complete your call, otherwise the chances of it getting done and keeping all the important thoughts 'top of mind' are very slim.

** Give a copy of your WIL to the office manager or doctor to discuss your goals.

**Date August 9 **Name ____________

**Practice Name ____________ **Phone ( ) ____________

**Mailing Address ____________

1. What I learned during my call, SPECIFICALLY.

- Take control of the call by asking for their name in the beginning of the call
- Have more of a sales mind-set

2. How I'm going to APPLY what I learned, SPECIFICALLY.

- "I'd be happy to help you w that, may I have your name please." → Before anything else.
- self-monitor your calls to improve your sales mind-set

3 Action Steps Following Each Monthly Teleseminar

1. **Email your WIL to:** support4.chrismullins@bluetie.com
2. Or, you can Fax 1-866-694-2563 - **please "print" LARGE and clear**
3. Give a copy of your WIL to the Office Manager or Doctor to discuss your individual goals.

What I Learned (WIL) Cheat Sheet

WIL's are the perfect way to begin to hold yourself accountable *following* your 'monthly group teleseminar' with Chris Mullins. For optimum results stay focused on the specific action steps you'll apply. **You do NOT need to use this form in order to submit your WIL; however you must answer the TWO questions below and provide ALL the information below. ☺

It's important that you discipline yourself to do your WIL once you complete your call, otherwise the chances of it getting done and keeping all the important thoughts 'top of mind' are very slim.

**** Give a copy of your WIL to the office manager or doctor to discuss your goals.**

**Date Aug 15 07 **Name Adriana

**Practice Name ____________ **Phone ( )

**Mailing Address ____________

1. What I learned during my call, SPECIFICALLY.

I learned that the most crucial tool in the practice is our script

2. How I'm going to APPLY what I learned, SPECIFICALLY.

making sure it's ALWAYS handy

July 31, 2007

Sandy

1. What I learned during my call, SPECIFICALLY.

Take control of the call right away. Build a relationship with the patient, if patient is not sure they want to make the appointment and still doing research, still tell them let me be the first one to welcome you to the office that changes there mind set into making the appointment, we need to stand out in there minds so they call us back. Document what the patient says word for word regarding there problem and then repeat it back to them exactly the way they said it. It shows them that we are listening, also other staff members can use there language when they come in for there appointment. Answer only questions patient ask you. Try not to give extra information if patient does not request it. Try not to pause to often, ask the patient for there name right away.

2. How I'm going to APPLY what I learned, SPECIFICALLY.

I will start documenting patients word for word on the computer so my team can repeat the problem to them when they come in for there appointment I like that idea it does feel like we are listening to them and we are here for them. I try not to give out to much info unless the patient request the information and still I don't give that much when they do request it I try to be short and sweet I don't want to overwhelm the patient.

1. **Email your WIL to:** support4.chrismullins@bluetie.com
2. Or, you can Fax 1-866-694-2563 - **please "print" LARGE and clear**
3. Give a copy of your WIL to the Office Manager or Doctor to discuss your individual goals.

What I Learned (WIL) Cheat Sheet

WIL's are the perfect way to begin to hold yourself accountable ***following*** **your 'monthly group teleseminar' with Chris Mullins. For optimum results stay focused on the specific action steps you'll apply.** **<u>You do NOT need to use this form in order to submit your WIL; however you must answer the TWO questions below and provide ALL the information below</u>. ☺

It's important that you <u>discipline</u> yourself to do your WIL once you complete your call, otherwise the chances of it getting done and keeping all the important thoughts 'top of mind' are very slim.

** **<u>Give a copy of your WIL to the office manager or doctor to discuss your goals.</u>**

**Date Aug 15/03 **Name Sandy

**Practice Name ______________ **Phone ______________

**Mailing Address ______________

1. What I learned during my call, <u>SPECIFICALLY</u>.

Scripts are the most important tool. Scripts show's you how to control the conversation. Mandatory a team member has a script in front of them in good condition, Hold staff accountable - Role playing and pop quizes with team Members

2. How I'm going to <u>APPLY</u> what I learned, <u>SPECIFICALLY</u>.

Our scripts are placed and laminated on the phone so you pick up the reciever its right there in front of you so there's no excuses. I will start holding pop quizes and start roleplaying with my team Members. once a week at least. Liz say's she will help in the team's pop quizes and role playing, will set her mind in appt making mode everything

3 Action Steps Following Each Monthly Teleseminar

1. **Email your WIL to:** support4.chrismullins@bluetie.com
2. Or, you can Fax 1-866-694-2563 - **please "print" LARGE and clear**
3. Give a copy of your WIL to the Office Manager or Doctor to discuss your individual goals.

What I Learned (WIL) Cheat Sheet

WIL's are the perfect way to begin to hold yourself accountable *following* your 'monthly group teleseminar' with Chris Mullins. For optimum results stay focused on the specific action steps you'll apply. **<u>You do NOT need to use this form in order to submit your WIL; however you must answer the TWO questions below and provide ALL the information below.</u> ☺

It's important that you <u>discipline</u> yourself to do your WIL once you complete your call, otherwise the chances of it getting done and keeping all the important thoughts 'top of mind' are very slim.

**** <u>Give a copy of your WIL to the office manager or doctor to discuss your goals.</u>**

**Date AUG 28/09 *Name LIZ

**Practice Name ______ **Phone ( )

**Mailing Address ______

1. What I learned during my call, <u>SPECIFICALLY</u>.

(1) PREPARE SCRIPT, USE GREETING AND PROPER TONE OF VOICE, GET NAME OF CALLER (VERY IMPORTANT)

(3) CONTROL THE CALL, LISTEN AND IDENTIFY THE CALLERS PROBLEM
- WRITE DOWN WHAT CALLER IS SAYING
- ASSURE CALLER THAT THEY MADE THE RIGHT DECISION
- REPEAT PROBLEM TO CALLER * SCHEDULE APPOINTMENT AS SOON AS POSSIBL

2. How I'm going to <u>APPLY</u> what I learned, <u>SPECIFICALLY</u>.

(1) IF PATIENT IS NOT READY TO SCHEDULE APPOINTMENT REPEAT STEP (5) (WHAT GOT YOU TO CALL US TODAY) AND TRY TO SCHEDULE AN APPOINTMENT OFFER A FREE CONSULT

(2) IF PATIENT IS STILL NOT READY COLLECT CONTACT INFORMATION AND SEND THEM AN OFFICE NEWS LETTER.

More Resources to Check Out

- **www.BigCaseMarketing.com**
- **www.GreatBottomLine.com** — These teleclinics include informal, no nonsense, fast-paced, how to, business, telephone skills, staff, sales and management tips to get your mental endurance batteries recharged for the rest of the week. Go to this web site to learn more about scripts and Chris Mullins, The Phone Sales Doctor™ and Phone Script Doctor™.
- **www.MullinsMediaGroup.com** — Done-For-You Call Tracking Solutions™: It's critical that you know your ROI on all your marketing efforts. Now, you can easily track all marketing results, record all inbound calls, review calls for quality assurance and to train the team on how to improve closing appointments, daily. It doesn't get any easier than this! The next time you drop a new marketing piece you'll know how many calls you received and you can listen to every call. The next step is to train your team how to handle calls differently which will absolutely improve the bottom line. Go to this web site to find out more about our Done-For-You Call Tracking Program.
- **www.PetethePrinter.com**
- **www.NoBSBooks.com**
- **www.BusinessTalkRadio.com**
- **www.Milteer.com**
- **www.GreatVoice.com**
- **www.DanKennedy.com**
- **www.Accuscreen.com**
- **www.ChrisMullinsConferencing.com**

Inspect What You Expect!

- **www.UniversityOfPhoneProfessionals.com**
- **Mystery shop calls playing prospect**
- **Monitor and critique real customer calls**
- **Done-for-You staff phone success training and coaching**
- **Monitoring for script adherence**
- **Mystery shopping live chat operators**

The Most Incredible **FREE Gift Ever**
The Critical In-Bound Call Training & Skills
$1184.94 Of Ongoing "How To" Phone Success Information

☐ I want to *test drive* Chris Mullins' MOST INCREDIBLE FREE GIFT EVER and receive a steady stream of The Phone Success Program™ with Chris Mullins and profit-making information which includes…

- **NEW! Limited Time Charter offer! Bi-monthly Mystery Shopper Call to your office and critique (Value – $397)**
- **Phone Success Program™ Gold Membership (Two Month Value – $99.94)**
  - o www.UniversityOfPhoneProfessionals.com Gold Member RESTRICTED ACCESS WEBSITE
  - o At least a 40% DISCOUNT to take OnlineTraining Center Phone Success Certification Self-paced Programs
  - o Two issues of *Chris Mullins Nuggets® for Sales* print newsletter
  - o Two issues of the TELEPHONE SKILLS CHEAT SHEET™
  - o Special FREE Monthly Question & Answer CALL-IN TIMES with Chris Mullins – NO QUESTION OFF LIMITS
- **The FAST START: Introduction to "Your Telephone – The LIFELINE to YOUR Business" how-to online VIDEO (Value – $397)**
- **Three Off-the-Cuff Teleclinics on how to "Improve Your Bottom Line With the Telephone" online audio presentations by Chris Mullins (Value – $291)**

There is a one-time charge of $19.95 in North America or $39.95 International to cover postage for 2 issues of the FREE Phone Success Program Gold Membership. You will automatically continue at the *lowest* Gold Member price of $49.97 per month. Should you decide to cancel your membership, you can do so at any time by calling Mullins Media Group at 603-924-1640 or faxing a cancellation note to 603-924-5770. Remember, your credit card will NOT be charged the low monthly membership fee until the beginning of the third month, which means you will have received 2 full issues to read, test and **profit from all the powerful techniques and strategies you get from being a Phone Success Member**. And, of course, it's impossible for you to lose, because if you don't absolutely LOVE everything you get, you can simply cancel your membership after the second free issue and never get billed a single penny for membership.

***** *E-mail is required in order to notify you about the Fast Start how-to online video that you will be invited to along with the Off-the-Cuff Teleclinics.***

Name: ______________________ Business Name: ______________________

Address: ____

City: ________________ State/Province: __________ Zip/Postal Code: __________

*E-mail: ____

Phone: ______________________ Fax: ______________________

Credit Card: ☐ Visa ☐ MasterCard ☐ AmEx

Credit card number: ______________________ Exp. Date: ______________________

Signature: ______________________ Date: ______________________

(Providing this information constitutes your permission for Mullins Media Group™ LLC to contact you regarding related information via mail, e-mail, fax and phone.)

Private, secured fax back to **(603) 924-5770**

Or mail to **Mullins Media Group LLC**

507 Greenfield Rd., Peterborough, NH 03458

AHA Notes: